A-Z
ATOMIC PHYSICS

A-Z ATOMIC PHYSICS

Prof. G.K. Bose

CENTRUM PRESS
NEW DELHI-110002 (INDIA)

CENTRUM PRESS
H.O.: 4360/4, Ansari Road, Daryaganj,
New Delhi-110 002 (India)
Ph.: 23278000, 23261597
B.O.: No. 1015, Ist Main Road, BSK IIIrd Stage
IIIrd Phase, IIIrd Block,
Bangalore - 560 085 (India)
Tel.: 080-41723429
Visit us at: www.centrumpress.com

A-Z Atomic Physics

First Edition, 2009
ISBN 978-93-80106-04-5

PRINTED IN INDIA

Printed at Salasar Imaging Systems, Delhi-110035 (India)

Contents

Preface

With several Nobel Prizes to its credit over the past decade, research in atomic, molecular and optical (AMO) science and engineering finds itself at the apex of investigations ranging from fundamental studies in quantum mechanics to the development of new nano-technologies. Whereas in years past, most researchers working in AMO fields had strong backgrounds in physics, optics and electromagnetic theory, this is not always the case today as the field and its participants become more and more diverse. Consequently, there is a need for a text designed to provide an overview of the field to get researchers new to AMO areas up and running as quickly as possible. This two-volume series entitled *Light-Matter Interactions* is designed to meet this need and to be user-friendly to investigators from physics, chemistry and the engineering disciplines.

Topics in atomic structure include the relativistic theory for highly charged atomic ions and calculations of parity nonconservation. Topics in atomic photoionization include single and double photoionization processes, and photoelectron angular distributions. Topics in electron-atom collisions include the theory of electron impact ionization, perturbation series methods, target dependence of the triply differential cross section, Thomas processes, R-matrix theory, close coupling, and distorted-wave theory.

Author

Chapter 1

Classical Mechanics

MACROPHYSICS AND MICROPHYSICS

Modern physics accepts the atomic structure of matter. Matter, palpable and visible to us, is made up of large numbers of minute bodies, called atoms, which can neither be seen nor felt. Greek philosophers had suspected the existence of atoms, and this concept taken over from the Greeks has played an important part in scientific investigation and development since the beginning of western scientific research. But it was not until this century that what had been supposition, speculation or fantasy was raised to the rank of solidly established scientific knowledge confirmed by direct experimental proof. It can even be said that ours is the century of atomic research in physics; and a survey of the ideas of modern physics should first of all deal with the investigation of atoms.

The moment, let us consider a piece of homogeneous matter capable of being subdivided. A piece of copper or a piece of rock crystal can be broken and split up into smaller and smaller pieces, and the fragments continue to show the characteristic properties of the copper or the quartz. Even if certain limits existed for the practical accomplishment of such a subdivision — it is difficult to break up into smaller parts quartz which has already been ground up to dust — yet this rough experiment presents no criterion for the existence of a definite, clear limit to further subdivision. The dust particles still differ in size and shape among themselves. Thus one might conclude, that — even though our tools are ineffective —

nature presents no obstacle to halving the smallest dust particle once more, and to halving this half again, and so on. . . without a limit.

Today we know with certainty that that is false. But very refined methods are required to attain the limit to the continued division of matter — in other words, for the experimental determination of the atomic structure of matter. But, as previously stated, such methods have been available for only a few years.

Thus there are large and rich fields of physical research in which the atomic structure of matter is not at all recognizable. Naturally no information about atoms can be expected from such research, but then no knowledge of atoms is necessary, as long as investigation is limited to these fields. It is customary to speak of macroscopic physics (or macrophysics) when referring to those investigations in which the presence of atoms is not discernible; on the other hand research which penetrates into the atomic detail of matter is denoted as microphysics.

This review is devoted in large part to a discussion of microphysics — but since only a thorough comprehension of macrophysics can make possible an understanding of microphysics, we shall begin with a discussion of the former. Moreover, it was in our century that macrophysics reached its final ma turity.

It was in these recent developments that certain philosophical theoretical thought processes were developed and clarified, making it possible to reconcile the surprising and paradoxical results of microphysical experiments, which previously were difficult to understand.

We are beginning with very simple things. Despite their simplicity, however, they merit careful and thoughtful consideration.

THE LAWS OF FALLING BODIES

The Greeks had already established the laws of statics, of mechanical equilibrium of levers and of fluids. The essentials of these laws had been clarified. On the other hand, dynamics, the theory of the mechanical laws of motion, was first established by Galileo. It is not within the province of this review to trace the historical development of Galileo's investigations; although for a deeper spiritual historical understanding of modern science, closer consideration of the historical origin of modern physical thought — up to the end of the Middles Ages — would be very valuable. We shall simply consider Galileo's laws of falling bodies in their finished form. The simplicity of these laws and the ease with which they can be interpreted today should not deceive us as to how tremendous was his mental capacity. To judge his contributions fairly, the historical-cultural background and the ideas — so completely changed and readjusted since then — prevalent in pre-Galilean times should be considered.

Let us consider the motion of a falling or pro jected body. Simple, everyday experience indicates that a feather falls more slowly than a piece of lead; and Aristotle had taught generally that light bodies fall more slowly than heavy ones. Galileo's main contribution is that he recognized the possibility of extracting an obscure but simple law from the amazingly complicated motions of real projected bodies; he considered falling motion in a vacuum. It is possible to prove that in an evacuated tube a feather falls as quickly as a piece of lead — this experiment is part of the course of study in schools today. But Galileo's contemporaries still considered the idea of a vacuum as impossible, as nonsense.

In those times it required far greater strength of intellectual abstraction than we might expect today, to recognize that it was possible to separate (mentally) all the associated events from the motion of falling bodies. We know today that the effects which originate from atmospheric pressure and the influence of the wind can lead the motion of a falling feather to proceed entirely differently from that of a falling stone.But it is possible to extract from all these complex and varied motions an ideal form which is to be regarded as the motion of a body in a vacuum. This "ideal" projectile motion is wonderfully simple compared to the actual motions of falling bodies or projectiles:

- Ideal projectile motion is always a plane motion (proceeding within a fixed plane perpendicular to the earth's surface).
- It proceeds independently of the mass, size and shape of the projected body; it simply depends upon the magnitude and direction of the initial velocity.

It is not influenced by any rotation of the projectile around its own centre of gravity; even if the body rotates around it, this centre of gravity moves according to the law of ideal projectile motion.These properties do not properly describe the motion of a real bullet, which is not exactly a plane motion, and which depends very substantially on the mass, size and shape of the projectile as well as on its spin (rotation).

Here we witness one of the historically most important examples of physical natural scientific thought. The phenomena presented to us directly by nature are so varied and complex that it is impossible for our narrow minds to grasp them individually. We must mentally divide these natural occurrences into their simpler components; we must consider results of artificially produced extremely unnatural conditions instead of those offered through direct experience. Thus we uncover the ideal cases, which can be subjected to more precise, considered treatment. These results in turn establish criteria for the evaluation of the real events, in each case different from the ideal to a greater or lesser extent.

Now let us consider the laws of ideal projectiles more exactly. For this we want to view the simplest case, namely vertical rectilinear fall. The body, first retained at rest and then released, falls downwards — as long as it continues to fall with ever increasing velocity. At the instant of release, the velocity is exactly zero, but thereafter it increases steadily. This increase does not proceed in intermittent, rapidly repeating, sudden jumps, but rather in an uninterrupted, continued growth, in which there are neither pauses nor sudden discontinuities. Mathematicians express it very clearly when they say that the velocity increases uniformly.

In such a case, when the velocity is changing uniformly and constantly so that it is never the same at two different points in time, no matter how closely they may follow one another, what does velocity mean?

When a body moves with uniform velocity it is clear what velocity means — the distance traversed is divided by the time of travel. We commonly say that a body traveled at a velocity of one meter per second or that an auto traveled at seventy kilometers per hour. But how are we to understand and define velocity abstractly if it doesn't remain fixed throughout even a fraction of a second?

In Galileo's times this question presented very great difficulty; the abstract and mathematical tools for solving it were still lacking. In an evaluation of Galileo's contributions, this fact, too, must be remembered.

Galileo had been able to dispose of these questions completely for ideal projectile motion; it only became evident later how great an achievement this was. Newton cleared up the concept of velocity (and acceleration) quite generally, for any motions. For this purpose he had to establish an entirely new branch of mathematics, the so-called differential calculus.

Differential calculus is undoubtedly the greatest creation achieved by western mathematicians. In this Leibnitz stands beside Newton as the founder — although his considerations did not emanate from mechanics, but rather from geometry. Naturally both of them, Leibnitz and Newton, referred back to forerunners and trailblazers; in the problems of their time and in their natural scientific and mathematical investigation there lay an inevitable compulsion toward this direction of thought. But what they produced by bold statements occupied generations of mathematicians after them and stimulated further great results.

This new mathematics, which is an original creation of western thought, not anticipated by the Greeks, deals with precisely those questions which confront us when we inquire about the exact meaning of the "concept" of velocity in the case of constantly changing velocity. The function of differential calculus is to express clearly mathematically quantities that describe constant, fluid, continuous change.

The exact definition of velocity can be expressed as follows. First, it is clear what is meant by the average velocity within a given time interval — the distance traveled in this time is divided by the length of time. Now, in order to obtain the exact velocity for a given point in time we choose a small time interval (including this point in time) and determine the average velocity associated with this interval; it will, if the time interval chosen is small enough, give an approximately correct value for the exact velocity desired. This value can be improved by replacing the time interval with one half as large; and then this improvement can be repeated. It is frequently possible to repeat it mentally at pleasure, and thus with unlimited approximations to approach ever more exactly the precise value desired.

This, therefore, is the definition of the concept of "velocity". The problem is no simpler if one wants to obtain the results quantitatively, and for this purpose wants to sharpen the concept so that it will make a mathematical evaluation possible -when one does not want to be content with a solely emotional, indefinite application of the concept. But one need not fear that this has led to a hopeless, practically insoluble problem in calculation. On the contrary, there are ingenious considerations (and their cultivation is the inherent content of differential calculus) which make it possible to establish in a specific form of motion the limiting value which we must take as the basis for the definition of velocity.

Galileo had already answered these questions for ideal falling motion. He had recognized that for this ideal motion, dissociated from air resistance and all secondary influences, a wonderfully simple law exists; namely, velocity increases proportionally with time. Thus, after double, triple, etc., time of fall it is exactly doubled, tripled, etc.

This law extends also to general ideal projectile motion. If we throw a body vertically upwards or downwards, its downward velocity will still always increase in proportion to time — which in the case of upward motion naturally means a corresponding decrease in the upward velocity. If it is thrown obliquely, the body's height measured vertically to the earth's surface (in ideal projectile motion) changes with the passage of time exactly as if it also were a vertical fall. Simultaneously, the horizontal distance from the starting point increases proportionally with time; or expressed differently, in the horizontal direction the motion proceeds with fixed velocity. Mathematical consideration of these determinations yields the fact that the projectile path (trajectory) is a parabola.

FORCE AND MOTION

The laws of generalized ideal projectile motion formed another example of how scientific thought made natural processes comprehensible through resolution into simpler components. We conceive of parabolic motion as simultaneous execution of two different motions, vertical and horizontal. It is also possible to study the horizontal motion isolated from the vertical -for this all we need is a horizontal rail, or perhaps a plane ice surface, on which the body can slide. Then for the ideal limiting case (i.e., in the case of not only the absence of air resistance, but also of any friction accompanying the sliding) we reach the conclusion that the body moves in the horizontal direction with constant velocity.

Galileo accepted the certainty of the spherical shape of the earth, at that time not a very old concept; and he could represent the earth as moving freely through space on the basis of the Copernican theory which he accepted. Thus he knew that the motions horizontal and vertical or above and below are only relative. And from this the possibility of generally recognizing the basic form of motion, free from external influences, in the ideal, resistancefree motion of a horizontally sliding body became evident. Newton, the first one to state this clearly, expressed his famous law of inertia as follows: a body moving in a vacuum unobstructed and free from external forces moves in a straight line with constant velocity.

Falling motion must thus be conceived as a deviation from the invariable behaviour of a body, for which the cause is to be sought in an attractive force emanating from the earth. Newton clarified for all cases the manner in which a force acting on its centre of gravity alters the motion of a body -the force manifests iteslf in a change of velocity, an acceleration. In vertical free fall the acceleration is constant — whereby we define acceleration as the increase in velocity divided by the duration of the fall.

For complex motion, in which the acceleration is no longer constant but is subjected to continuous changes the same considerations are necessary as were introduced for the concept of velocity. Newton's new mathematics, differential calculus, mastered all these problems with one stroke.

If we limit ourselves to rectilinear, uni-dimensional motion, then according to Newton the acceleration is exactly equal to the force divided by the mass of the accelerated body. It depends on neither size nor shape, nor material, colour nor temperature of the body, but on its mass alone.

The laws of three-dimensional motion are analogously simple and general. Here it is simply neces sary to resolve the motion into three components, acting in mutually perpendicular directions, whose simultaneous execution yields the total effective motion. If the force acting on the body's centre of gravity is resolved correspondingly, the above Newtonian law is valid for each of the three uni-dimensional portions of the whole process.

This knowledge made possible a clear definition of the concept of physical causality. The general notion that nothing happens unless a definite cause exists for it was elevated to a quantitative law: a definite force acting on a body imparts a definite acceleration to that body.

This point merits further consideration. To really understand the significance and importance of the ideas which constitute physics, we must always be ready to refer back to previous concepts. Only in this way can we realise how revolutionary such thoughts were at one time, although they have become quite familiar and conversant to us.

As the most prominent feature of natural-physical research we recognize the search for quantitative, mathematically comprehensible laws. We must remember that extreme idealization of natural events is necessary to render this search successful. Since, by eliminating air resistance,

Galileo attained the process of ideal fall, he was able to uncover mathematically simple, beautiful, exact laws; and conversely, these laws show that he looked for the idealization in the proper direction.

Naturally calculations on the projectile trajectories of modern weapons do not yield these ideal projection laws at all. Such calculations require consideration of air resistance — which also is introduced in a more or less idealized form. Ballistics, which deals with these problems, is a special, highly developed science whose outstanding practical significance needs no special emphasis. But the science of ballistics had nothing new to add for the general development of physical science, for the development of physical thinking and ideas.

The idea of continuity, which attained its mathematical form in differential calculus, is important for the clear understanding of motive processes. We also want to make it clear immediately that this continuity of natural events — "natura non facit saltus" — was already evident in the elementary fact that it was at all possible to speak of a definite trajectory of a moving body.

A body cannot reach one place from another by jerks, suddenly disappearing here and emerging again there; it must describe a continuous connected path between the two. But why is that necessary? We know from experience that it always is that way, but is there a logical necessity that it cannot be otherwise? These questions are not idly posed: we shall never be able to understand microphysics unless we have carefully examined such questions.

If we continue to consider macrophysics, it must be realized that in this sphere the principle of continuity is valid without exception. When a shell explodes, the proposition holds that individual fragments can only change their positions through continuous motion, rapid as it may be. Or in an automobile accident — the law obtains that each body can only change its velocity continuously, never with "discontinuous", complete suddenness.

Also, the quantitative, mathematical definition of mechanical causality, in which Galileo's and Newton's knowledge culminated, is inseparably bound up with the concept of continuity. For, it is through Newton's law — "force equals mass times acceleration" — that constant changes in velocity are traced back to the forces which cause them.

RELATIVE MOTION

The knowledge expressed in the law of inertia and in Newton's definition of the operation of force was diametrically opposed to earlier views. Previously the problem had been viewed just in reverse — explanations were being sought for the fact that a hurled stone retains its velocity after being released from the hand. It was considered natural and understandable that it must lose its velocity if the impulse is lacking to maintain it. For everyday experience does indicate that a wagon, for example, which is supposed to move uniformly along a straight path, does require a continuous uniform force from the beast drawing it.

The now recognized fact that, exactly opposite, an uninfluenced body retains its velocity without any change — the wagon on the road is not a valid example of this because it is retarded by the action of friction — and that quite generally only acceleration, and not velocity, is determined directly by the acting force, is related to certain very significant problems, which were not fully solved until this century. For the moment these problems are pertinent only so far as they refer to the mechanics of Galileo and Newton.

Opponents of the Copernican theory of the motion of the earth around the sun had introduced the following objection: if a gun is shot vertically upward, the shot falls down close to the gun. But, they said, according to Copernicus, since the earth has in the meantime moved along some distance, the shot should have fallen down in an entirely different place. This was impressively demonstrated in an experiment by Gassendi: he dropped a stone from the tip of the mast of a rapidly traveling boat. It fell down below near the mast, not on the stern of the boat nor behind the boat in the water.

Galilean-Newtonian mechanics explained this result without more ado. And, since generally it predicted suitable results for all analogous experiments, it lent an indispensable contribution to the justification of the Copernican theory. To our present thoughts, which accept the ideas of Galileo, Newton and their associated mechanics as a basis, the process is clear immediately: the stone, first held in place at the tip of the mast and then released, received the same velocity in the horizontal direction as the boat; and it retained this horizontal velocity (as long as the effect of air resistance was negligible) so that as it fell downwards it moved along with the ship in the horizontal direction.

From the Newtonian laws of mechanics it is obvious that if a boat (or train) is traveling in a straight line with constant velocity, one cannot determine on the inside of a closed cabin of the ship whether or how fast the boat (train) is traveling. Although naturally it can be deduced from the altered rocking of the boat or swaying of the train, here we are considering the ideal case of a vehicle moving along smoothly without rocking or bumping.

An object dropped in the cabin will fall vertically to the floor, precisely as it obeys the law of falling bodies on land. Now let us consider that an observer standing on shore can see inside of the cabin. The author desires to establish that the falling object is still subject to the laws of falling bodies, although this observer would not see it fall vertically, but in a parabola — with a horizontal velocity (as long as the effect of air resistance was that of the boat).

The concept of velocity is relative. The observer on shore views all objects in the cabin with a different velocity than the experimenting traveller; for the witness on land every object moving in the cabin acquires in its motion also the additional velocity of the vessel. Therefore, to avoid misunderstanding we must always append to the word velocity "relative to the boat" or "relative to land".

When a boat travels with constant velocity (relative to shore) then the acceleration of any body relative to the ship is always the same as it is relative to land despite the difference between the relative velocities. Consequently the same mechanical laws apply to motion relative to the boat and relative to land. In both reference systems the same accelerations are caused by the forces acting.

This discovery, that one can never determine uniform motion of a closed room from its inside with any mechanical apparatus — be it simple or complex — whose method of operation is subject to Newton's laws, is called the principle of relativity. The principle of relativity is, beside the principle of the conservation of energy, the most general and most comprehensive physical law that we know. As yet we have only established its validity with respect to the laws of Galilean-Newtonian mechanics. But we shall see that its significance extends far beyond this and embraces all fields of physics.

Therefore we can't perceive through any mechanical experiments conducted on the earth that the earth (relative to the sun) glides along thirty kilometers per second, or that the sun (relative to the milky way system) flies along with still greater velocity. And we can't distinguish whether the milky way, which despite its tremendous size is only a small island in space, for its part executes as a unit any uniform rectilinear motion.

The rotation of a body, for example, can be recognized by means of mechanical effects. Anyone sitting on a carousel can tell, even with his eyes closed, that it is turning from the centrifugal force which presses his body against the outer wall — since according to the law of inertia the body would fly out in a tangential direction due to the rotary motion if the outer wall didn't retain it. In the same way we notice a curve within a train; we also notice the acceleration or deceleration of rectilinear motion when starting or stopping. It is only uniform progression which is not noticed, because only there is the acceleration of the body relative to the vehicle the same as that relative to the fixed rails.

In the case of our earth the centrifugal force caused by its rotation is evidenced in the flattening of the poles. It is also possible to measure the earth's rotation directly through a simple mechanical experiment — whereby the Copernican conception of motion in our solar system, developed from astronomical knowledge, was shown to be a necessary result of Newtonian mechanics. This is the famous Foucault pendulum experiment.

A weight swinging on a very long thread transferred into an elliptical form of vibration under the influence of the earth's rotation. The measurement of this effect permits the determination, based on Newtonian mechanics, that the earth actually rotates with the same speed that the "fixed star" sky seems to move (oppositely directed); that actually the "fixed star" sky, as Copernicus taught, remains at rest.

NEWTON'S LAW OF GRAVITATION

The most significant addition that Newton made to Galileo's knowledge was his teaching that the motions of the planets, conceived in the Copernican sense, were to be considered as consequences of the same mechanical laws which were derived from terrestrial falling motion. Newton availed himself of the opportunity to demonstrate conclusively his extension of Galileo's ideas, using as an example the most wonderful mechanical system provided by nature.

To his admiring contemporaries the explanation of known processes made possible by the new mechanical principles was convincing proof of the validity and profundity of these principles. It was not exaggeration when H. Poincaré once said that mankind had learned mechanics from celestial and planetary motions.

To his general mechanical laws, valid for all effects of mechanical force, Newton added the law of gravitation: any two bodies exert an attractive force on each other that is proportional to the product of the two masses (thus, is doubled, tripled . . . with the doubling, tripling . . . of either mass), and that becomes weaker as the separation between the two bodies is increased. The decrease in attractive force with distance is such that, e.g., the attractive force of the sun on a meteoric stone in interstellar space decreases with increasing separation from the sun in the same manner as the strength of the light emanating from the sun, which becomes increasingly thinned out with increasing separation. Mathematically expressed, the force of gravitation varies inversely as the square of the distance. [1] Newton, by mathematical reasoning, showed that the laws of the motions of the planets (discovered by Kepler) are a necessary result of the gravitational attraction between the sun and the planets. The Keplerian laws state:

- The orbit of each planet is an ellipse with the sun at one of its foci.
- Each planet revolves so that a line joining it to the sun sweeps over equal areas in equal intervals of time — so that when the planet is closer to the sun it moves more rapidly than when it is at a more distant point in its orbit.
- The squares of the periods of any two plan ets are in the same proportion as the cubes of their mean distance from the sun.

It is obvious that Kepler's laws differ in nature from the mechanical laws which formed the basis for their final mathematical proof. Actually, Kepler himself viewed his laws in a manner different from Newton's. For Kepler, his laws were primarily an expression of the beauty and harmony of the divine creation; whereas Newton's laws demonstrated in the planetary system the principle of causality in nature, according to which the resultant motions proceed necessarily from the acting forces.

This same gravitational force which regulates planetary motions also determines the orbit of our moon around the earth, or of her moons around Jupiter. This same force which controls the solar system with its planets, comets and moons is also responsible for the fall of bodies on the earth.

In 1666 Newton had already visualized this comprehensive picture of the validity of his law of gravitation. But the calculations did not agree with observations. The magnitude of the earth's attraction working on the moon calculated from the motion of the moon did not appear to correspond in the sense of the Newtonian law with the magnitude of the force of gravity at the earth's surface. Newton concluded therefore, that still other forces come into play here; and unsatisfied, he set this investigation aside.

Not until sixteen years later, when he learned that new geodetic measurements of the circumference of the terrestrial globe had yielded a value about one-sixth greater than had been accepted until then, did he resume this work. Now Newton's calculations agreed with observations. Four years later, in 1686, his great work *"Philosophiì Naturalis Principia Mathematica"* appeared. In it he developed the principles of mechanics, set up the law of gravitation and taught the mechanics of the planetary systems to be understood on these bases; he also explained the phenomena of the tides.

Shells from modern guns that shoot beyond 100 kilometers have projectile paths whose calculation must include the consideration that the earth is not flat, but is a sphere.

Technology thus began to bridge the wide chasm which was opened by Newton's boldness of mind when he recognized that the motions of the moon around the earth and the planets around the sun are dependent upon the same laws as projectile motion on the earth's surface. Newton bridged this gap in the following hypothetical experiment: a projectile is shot off from a high mountain — somewhat horizontally. The experiment is performed repeatedly, with ever increased projectile velocity.

The projectile strikes ground farther and farther away; when the velocity is sufficient, at the other side of the earth; and with still greater velocity the case is reached wherein the projectile, by flying around the earth, again reaches the point from which it was shot off. In the "ideal" case of absence of air resistance -which condition actually prevails in interstellar space where the planets move around — the projectile will have maintained its initial velocity; consequently it will describe the same path again; and con tinuing, it will circle the earth permanently.

A characteristic feature of physical research can be detected in this Newtonian extension to planetary mechanics of Galileo's investigation of projectile motion. After obtaining a sure footing in a limited field an attempt is made to transcend the existing boundaries with the conceptions gained in the immediate field serving as a basis for the investigation of the more remote.

By trying to find connection and similarity between the newer phenomena and those previously investigated, an extension and generalization of ideas and recognized laws can be reached. Such generalized laws make it possible to "understand" the entire, extended sphere of experience.

We have just placed the word "understand" in quotation marks in order to emphasize, the importance and profundity of the problem of being clear about what is really meant by "understanding" physical events – and what goal is really striven towards in physical investigation and search for knowledge. Newton's celestial mechanics – probably on the whole the greatest contribution of physical thinking ever accomplished – indicated clearly that "understanding" means nothing else than tracing back the new to the already known. Aristotle's question, why a body that has been set in motion retains its velocity, was not answered, but simply brushed aside.

The law of inertia establishes that a body invariably and indefinitely retains its velocity as long as retarding forces do not act upon it. And once one is satisfied with this establishment -that is not to be "understood", but simply accepted – one also sees the problem of planetary motion from an entirely new angle. One need no longer (as Kepler still desired) assume in the sun the seat of permanent impulsion of planetary motion.

One other point merits our attention before we leave Newtonian mechanics for new subjects. What are the results of the planets' mutual attraction for one another? The mathematical solution of this problem is extremely involved and difficult (the famous "three body" problem). One fact is clear – since the sun's mass is tremendously large in comparison with the masses of the planets, the attractive forces exerted on each planet by the other ones are very weak in comparison with the attraction of the sun. Thus it is understandable that the Keplerian elliptical paths actually do prevail, with only small deviations, weak "perturbations", caused by the mutual effects of the planets on each other.

But shouldn't it be expected that these small disturbances could grow with the passing of time -in such a way that they might change the magnitude and positions of the elliptical orbits completely in the course of millions of years, and the planets would finally have to collide or plunge into the sun?

Newton entertained this suspicion. And he believed that from time to time the world creator must intervene to prevent such destruction. Therefore it was of great significance for the internal consolidation of the natural physical conception that Lagrange and Laplace later showed by clever mathematical proofs that the planetary system possesses "stability" of itself on the basis of the Newtonian law.

Therewith the idea was accepted that the paths of all events in the universe are controlled in their entire sequence without exception by the mathematically defined natural laws of physics.

Chapter 2

Model of Atoms

RUTHERFORD'S MODEL

The perception of an atom as the least indivisible particle of substance was the foundation of the cinematic gas theory. It have explained and predicted a series of the experimental phenomena.

However, at the end of XIX century new experimental data proved the composite structure of an atom. The experimentally found laws of radiation and uptake of energy by atoms (spectral series, radiation of an ideal black body, photoeffect) have not received an explanation within the framework of a classical electrodynamics. Only new ideas, based on the quantum theory, helped explaining these phenomenon.

Structure of Atom

British physicist Ernest Rutherford (1871-1937) has created a planetary model of atom exploring a dispersion of a-particles transiting through a thin metal foil. According to this model an atom looks like a tiny planetary system, in which the forces of an electrical attraction operate.

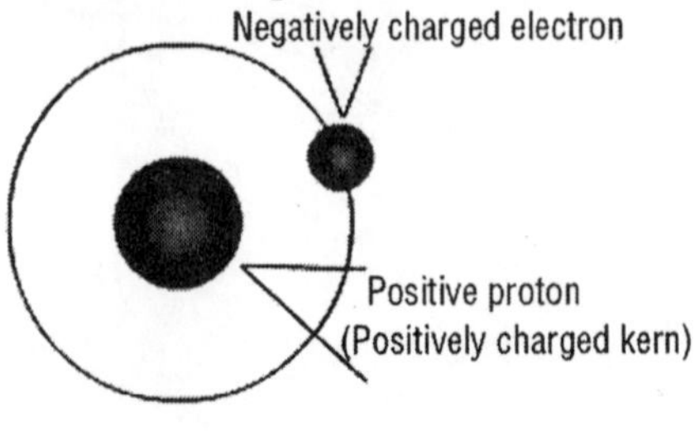

The centre of an atom is a positively charged kern. Almost all mass of an atom is focused there. Negatively charged eletrons rotate around the kern.

An atom of hydrogen has the most simple structure. Positively charged kern of an atom of hydrogen was called a proton. The only electron rotates around it. It was determined later, that every atom's kern consists of protons and neutrons. The neutron's mass is almost equal to proton's, but has not electric charge.

The mass of an electron is about 1/1840 of a proton's mass and does not influence an atomic weight of substance. The number of positive and negative charges of atom determines the position of a substance in a periodic system. The positive charge of an atom's kern is equal to a serial number of a substance in the table.

The number of neutrons in an atom's kern is equal to a residual between a rounded atomic weight and a number of positive protons in a kern (serial number of substance). The electrical neutrality rule of atom says, that the number of electrons rotating around the kern is equal to a number of positive protons contained in an atom.

Parameters of Atom

According to the Rutherford's experimental results the radius of a trajectory of an electron in a hydrogen atom is 0,53 x 10^{-10} m, the diameter of an atom of hydrogen is equal to 2,6 x 10^{-15} m. Volume of electrons and kern is approximately 1/10 000 of an atom's volume.

The substance of a kern has huge density of 200 million ton in 1 cm^3. It is all because the most of an atom's mass is concentrated in its kern.

The mass of a positive proton is approximately equal to the mass of a neutron and is 1,67 x 10^{-27}. The mass of an electron is 9,11 x 10^{-31} kg.

The classic radius of an electron, r_0, is equal to 2,8 x 10^{-17} m. It is the distance, on which the electrostatic energy of an electron intercourse with a point charge of the same quantity is equal to natural energy of an electron:

$$e^2 r_0 = m_0 c^2$$

e ¯elementary charge; c ¯speed of light in vacuum;
m_0 and nbsp- electronic mass

Deficiencies of the Theory

The Rutherford model of atom was well correlated with the results of a dispersion of an a-particle by atoms of substance, but it has not explained neither the process of radiation of atoms, nor legitimacies in spectrums of radiation. According to the laws of an electrodynamics rotating around the kern the electron should radiate electromagnetic waves. As a result of radiation the natural energy of an electron should be diminished, thus the trajectory it will be figured by a spiral, and during the order 10^{-8} about an electron should fall on a kern. Such deduction obtained on the basis of representations of classical physics about radiation, contradicted known stability of atoms and character of nuclear spectrums of radiation.

BOHR'S MODEL

Structure of Atom

The Danish physicist Niels Henrik David Bohr (1885-1962) has created the theory of a structure of atom, which was deprived of the basic deficiencies of Rutherford's model of atom. Thus he had to change some habitual representations of classical physics. Bohr has formulated three postulates (sometimes the first and the second postulates are combined, then Bohr's theory of atom is reduced to two postulates):

- The electrons in atom rotate around the kern on particular stationary orbits only. The motions on these orbits correspond to stationary states of atom, which do not vary in time without exterior actions. Rotating on stationary orbits, the electrons do not radiate energy.
- Allowed, discrete values of energy of an electron correspond to stationary states of atom.

This postulate was called the quantization rule of orbits. It affirms, that when the electron is going on a stationary orbit, it

Should have discrete values of a moment of motion:

$$m_0\, ur = n\, \frac{h}{2\pi},\ n = 1, 2, 3, \ldots$$

*m*0 and nbsp- electronic mass *u* - velocity of atom;
r- radius of an orbit *h*- Planck constant;

$$h = 6.62 \times 10^{-34}$$

- The radiation or uptake of energy by atom occurs only at transition of an electron from one stationary orbit to another. Thus quantum of light energy (photon) is radiated or is immersed. The energy of a quantum is equal to a difference of energies of an electron on the relevant stationary orbits: frequency of a light, energy of an electron on m and n orbits.

The laws of classical physics say, that the transition of bodies from one state to another occurs continuously at continuous radiation or uptake of energy. The classical laws have appeared inapplicable to the phenomena, which are described by Bohr.

Parameters of Atom

The radiuses of stationary orbits and energies of electrons are determined by dependence of energy of atom on radius of an electronic orbit.

The elementary system is the hydrogen-like atom. The hydrogen-like atom is the one, in which one electron rotates on a circular orbit around of a kern, which has a charge *Ze* (Z – serial number of a device in periodic system).

The energy of an electron in atom consists of a kinetic energy of a motion on an orbit and potential energy in an electric field of a kern (electronic mass, travelling speed of an electron on an orbit, radius of an orbit, permittivity of vacuum of empty space):

$$E = \frac{m_0 u^2}{2} + \frac{-Ze^2}{4\pi\varepsilon_0 r} = \frac{m_0 u^2}{2} - \frac{-Ze^2}{4\pi\varepsilon_0 r}$$

m_0 and nbsp- electronic mass; u - velocity of atorn
ε_0 -permittivity of vacuum of empty space
r- radius of an orbit

The kinetic energy is determined with the requirement, that a centripetal acceleration is created by force of a Coulomb attraction of charges. The kinetic energy of an electron is inversely proportional to radius of an orbit. A total energy:

$$E = \frac{1}{2}\frac{Ze^2}{4\pi\varepsilon_0 r} - \frac{-Ze^2}{4\pi\varepsilon_0 r} = -\frac{1}{2}\frac{Ze^2}{4\pi\varepsilon_0 r}$$

The total energy of an electron varies with the change of a radius of an orbit the same as potential energy does. Hence, the more than radius of an orbit of an electron, which rotates around the kern, the larger is the total energy of atom.

Value $r_1 = 0{,}528 \times 10^{-10}$ m is called the first Bohr's radius. For the subsequent values: $r_n = n^2 r_1$

n - main quantum number. It specifies an energy level, on which the atom is. The lowest state of atom is inconvertible. All atoms are in the lowest state at standard conditions. The atom can be exited, if the energy transferred to it. Thus it will be transferred to one of higher energy states. The distance from the lowest energy level $n = 1$ up to the proximate level with higher energy n = 2 makes 10,1 eV = 16,2 x 10-19 J for the atom of hydrogen. This energy is the least bundle of energy, which the atom of hydrogen can immerse, when it is in a normal state.

Deficiencies of the Theory

Bohr's theory explained the nature of a spectral series of hydrogen. It has allowed to spot theoretically the sizes of atom and energy levels of electrons for atom of hydrogen. Deficiency of Bohr's theory is the inconsistency of its build-up. Bohr's postulates have quantum character, but the stationary electronic orbits and allowed levels of energy of atom are determined by methods of a classical mechanics and electrodynamics. Therefore Bohr's theory has appeared usable to calculations only of monovalent atoms.

Bohr's theory does not give an opportunity to calculate intensity of spectral lines, which is interlinked to probabilities of electronic transitions. The introduction of quantum rules for definition of orbits and levels of electrons was not logically justified by Bohr.

The further development of the Bohr's theory was made by German physicist Arnold Johannes Wilhelm Sommerfeld (1868-1951). He has changed circular orbits on elliptic. It has enabled to apply Bohr's theory to multi electronic atoms. However the creation of a fundamental theory of atom is bound to the creation of a quantum mechanics.

MULTIELECTRONIC ATOMS

Structure of Atom

Just as in the atom of hydrogen, in more composite atoms electrons can rotate around of a kern on allowed orbits only. The orbits in composite atoms cluster in system of shells. Each shell contains a particular number of orbits, an electron can be on any og them. A *K*-shell is the closest to a kern and contains two orbits. The second is a *L*-shell has eight orbits. The third is a *M*-shell has eight orbits, the fourth *N*-shell has 18 orbits etc.

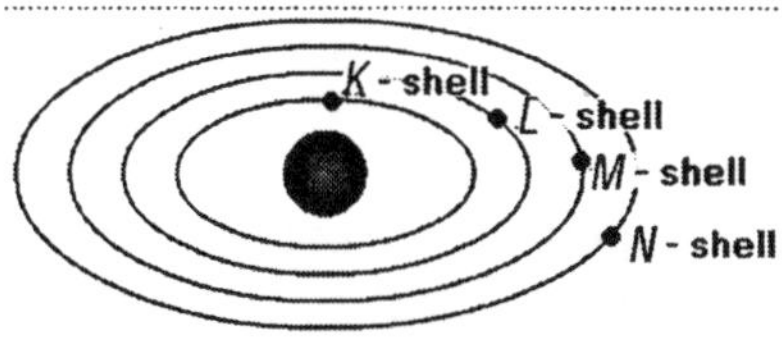

Parameters of Atom

When the electron travels from an orbit of greater radius to an orbit of smaller radius, the energy is oozed. But the transition from outer shells on interior is possible only when there are free orbits on the leter. The outer-shell electrons are less bound to a kern than interior ones: they are on greater distance from a kern, and the electrons, which are on inner shells, screen the nuclear charge. Therefore just a small amount of energy is sufficient for excitation of outer-shell electrons. The quantums of energy radiated at transition of such electrons in a nonexcited state have a quantity of about several electron-volt. Greater energy is necessary to detach inner-shell electrons from the atom. The greater the nuclear charge the greater energy is needed. Therefore, when there is a transition of electrons from outer shells to a *K*-shell in heavy atoms, the emission of quantums with energy of hundreds and even thousand electron-volt is possible. Such quantums of high energy are called Roentgen rays.

X-ray Spectrums of Atoms

The X-rays results from losses by atom of an inner-shell electron: The transition of electrons is accompanied by an emission of quantums of high energy with a small wave length, which are termed as Roentgen rays. The X-rays emerges when atom losses inner-shell electron: The transition of electrons is accompanied by an emission of quantums of high energy with a small wave length called Roentgen rays. The atom can lose inner-shell electrons as a result of an electron bombardment with electrons speeded up to energy of several thousand electron-volt. The wave length of the X-rays is comparable to interatomic distance in solid bodies.

The examination of the X-ray diffraction on crystal lattices of solid bodies enables, knowing a wave length of Roentgen rays, to determine interatomic distances in them. The examination of Roentgen rays also allows to determine nuclear charge, because it depends on a prime relation of an electron-binding energy in atom, which is expressed through energy of the X-rays. This relation is called Moseley's law:

$$v = A\,(Z - B)$$

v and -frequency of a line in a X-ray charateristic spectrum of a device

A and B - constants Z -serial number of a device

PROPERTIES OF LIGHT

In one phenomena, which are bound to distribution of light, its undular properties (interference, diffraction) are shown. In other phenomena its quantum properties (photoeffect, radiation and uptake of a light, process of interaction of a light with substance) are shown.

The light has simultaneously undular and quantum properties, which are differently shown in the different phenomena. However, when the wave length of a light is diminished, its undular properties are attenuated and the quantum properties are shown.

Compton's Effect

The experience of Arthur Holly Compton (1892-1962) on a dispersion of Roentgen rays by substances, which consist of atoms of light devices, have shown, that the dispelled Roentgen rays have a major wave length λ', than impinging λ. Thus the residual $\Delta\lambda = \lambda' - \lambda$ depends on properties of scattering substance and wave length of an impinging light:

$$\Delta\lambda = \lambda' - \lambda = 2\lambda_K \sin^2 \frac{\theta}{2}$$

The quantity λ_K is a stationary value for all substances also is termed as Compton's wave length, $\lambda_K = 2{,}43 \times 10^{12}$ m. This phenomenon, which is termed as Compton's recoil, is impossible to explain by a wave theory of light. In quantum theory the magnification of wave length of radiation at a dispersion is explained by interaction of photons of impinging radiation with electrons of atoms and molecules.

The part of energy of an impinging photon, which is transmitted to electrons of scattering substance, depends on a scattering angle of photon.

At $\theta = \pi$, that is provided that after a dispersion a photon will fly in the party, which is opposite to a tentative direction, the losses by a photon of energy will be maximal, that will give in maximal magnification of a wave length of radiation.

Light Pressure

Some properties, for example light pressure, can be surveyed simultaneously in the electromagnetic theory and in the quantum theory. According to the electromagnetic theory of light, the pressure p, which renders a luminous flux with power W, impinging normally on a simple surface, is equal:

$$p = \frac{W}{c}(1+R)$$

For an ideal black body (R = 0): $p = \frac{W}{c}$

For an absolute reflecting bodyh (R = 1) $p = \frac{2W}{c}$

If N of quantums of a luminous flux with frequency ν, which bears(carries) energy W = $N \cdot h\nu$, impinges on a simple surface for 1 second, then the blanket impulse at a reflectivity R will look like:

$$p = \frac{N \cdot h\nu}{c}(1+R) = \frac{W}{c}(1+R)$$

Thus, electromagnetic theory and quantum theory give identical results. However it is essential different representations will be utillized in the electromagnetic and quantum theories of light. The concept of an electromagnetic wave of application to a light guesses existence of a wave in particular field(area) of space, which is capable to contain even some lengths of waves of light.

The light quantums are considered as a particle, which are localized in very small volume of space. The representation about a light wave guesses a continuous distribution of energy in a spatial wave, the representation about quantums guesses localization of energy in a small part of space. Both these incompatible representations reflect unity of a wave-corpuscle nature of light, which is shown at distribution of light and its interaction with substance.

PROPERTIES OF SUBSTANCE

De Broglie's Theory

The French physicist Louis Victor duc de Broglie has applied wave-corpuscle treatment and to particles of substance: as the light detects corpuscular properties, also particles of substance should detect undular properties. De Broglie has utillized already wave-corpuscle properties, accepted for a light, and their quantum rule and has expressed a wave length through the performances of a particle. For a light the corpuscular point of view determines energy and impulse accordingly by expressions $E = mc^2$ and $p = mc$.

The wave length of a particle of substance is determined under the formula:

$$\lambda = \frac{h}{p} = \frac{h}{m\upsilon}$$

At adding in a rule of quantization of a moment of momentum of an electron the impulse of an electron, will turn out a relation $n\lambda = 2\pi r$. From it follows, that on length of a stationary orbit of an electron the integer of lengths of waves should be stacked. Thus, the requirement of existence of allowed orbits, which was injected by Bohr frequency relation, acquires legible physical sense: the allowed orbits are orbits supposing formation on them of standing waves of an electron. For a rectilinear motion of particles de Broglie has offered expression, which is made by analogy with the equation for a spread flat light wave in the complex shape:

$$\psi = \psi_0 e\ i(wt - \frac{x}{\lambda})$$

De Broglie's Waves

The specified waves have received a title of associated waves. Thus, the associated waves are traveling waves for freely propellent electrons or standing waves for electrons, which are bound in atoms. De Broglie's waves are not waves of propellent substance, they have not analog in classical physics.

De Broglie's guess about existence of undular properties of particles carries universal character: undular properties should have the electron, positive proton, neutron, atom, molecule, any moving object.

However objects, which have major masses and move with usual velocities, will have a so small wave length To in comparison with the sizes of objects, that the phenomena of an interference and diffraction for them can completely be neglected. The undular properties of a light are clearly shown in cases, when the wave length has compared to the sizes of bodies, with which the light interreacts.

The lengths of waves of electrons in usual requirements have the order of the nuclear sizes, hence, the effects are characteristic for them which are observed usually for Roentgen rays. For such small lengths of waves it is possible to observe a diffraction on nuclear crystal lattices.

EXAMINATION OF PROPERTIES

The existence of undular properties of substance was revealed at observation of patterns of a diffraction of beams of particles. Legible patterns of an interference and the diffractions give bundles of electrons, which have almost identical impulses. These patterns are very similar to optical patterns of an interference and diffraction.

At change of quantity of a momentum of electrons and measuring of a wave length, it is possible to receive a relation between these quantities. It has appeared equal to a known relation

$$\lambda = \frac{h}{p}$$

This de Broglie's relation is valid for any kind of particles. The undular properties of particles were detected at observation of a pattern of the diffraction of the bundle of electrons on chips. At measuring a scattering angle it is possible to spot a wave length of electrons.

Thus, at known quantity of potential, which was utilised for acceleration of a bundle of electrons, their velocity is possible to calculate, and then to receive independent value of a wave length of an electron. The value of a wave length of an electron, which was spotted in such a way, completely coincides with quantity of a wave length, which is found from a diffraction pattern.

The undular properties of electrons also were detected in experience with a cathode rays at major velocity, which was passed through a thin metal foil on a photographic plate. After development of a photoplate on it the diffraction pattern from rings of different intensity was visible. However this diffraction pattern was stipulated not by Roentgen rays, which could arise at a concussion of high-velocity electrons with metal. It confirms presence of a stream of charged particles.

MECHANICS

Quantum Mechanics is the study of matter and energy on the subatomic scale. It was came into being because classical physics could not explain certain experimental results, such as why the photoelectric effect occured only when the light shone is above a certain frequency. Quantum Mechanics and Relativity became the foundation of modern physics.

Main Points

Principle of Quantatization

The fundame ntal principle of Quantum Mechanics, and its namesake, is that energy came in discrete particles called "quanta". This idea, the Quantatization of Energy, was put forth by Planck in his Quantum Hypothesis(1900), an explanation of blackbody radiation. It was a departure from Classical Physics, which asserted that energy could be infinately small.

Photoelectric Effect

This quantatization of energy approach was used by Einstein in his 1905 explanation of the photoelectric effect, the emission of electrons by a metal when light is shone on it. No electrons are emitted when the light is below a threshold frequency, no matter how intense the light. Einstein postulated that light, while exibiting wave properties as shone by the Double-slit Experiment, also came in the from of particles called "photons", discrete packets of energy.

The energy of a photon is proportional to the frequency of the light(E=hv); thus, light with higher frequency had more energy and could excite the electrons, while light whose frequency were below the threshold could not, no matter how many photons there were.

Bohr's Model and Spectral Lines

Bohr also used the Quantatization of energy in his 1913 model of the atom. He stated that electrons revolved around the nucleus much like the planets revolve around the sun. They could be found in predefined, stationary orbits, radiating no energy as they revolve. Only when they gain a quatum of energy equivalent to the difference between the energy of the orbits do they become excited and enter a higher energy orbit. This is consistent with Einstein's explanation of the photoelectric effect, and explains why spectral lines only occur at certain places.

Wave-Particle Duality of Matter

In 1923 De Broglie expanded the Principle of Wave-particle Duality of Light put forth by Einstein to include all matter, called Principle of Wave-particle Duality of Matter. He asserted that all matter, not just light, exibit both wave-like and particle-like behaviour,

Quantum Matrix and Wave Mechanics

By now there was enough theories to put together a quantum picture of the atom. However, it was very difficult to sort through all the math. Heisenherg and Shroedinger simultaneously and independently worked to put forth a complete quantum mechanics. Heisenberg employed use of matrices, and his model was called the matrix mechanics. His approach emphasized the quantum—discrete properties while Schroedinger concentrated on the wave properties. He tried to find a function that would encompass all informatin about a particle or system at any given time. This is called the wave function. While debated ensued about which was correct, Schoedinger proved shortly after that the matrix mechanics and the wave function were mathematically equivalent.

Uncertainty Principle

Heisenberg then came up with the Uncertainty Principle, which stated that the position and velocity(momentum, which is mass*velocity) cannot be known precisely simultaneouely.

The more accurately one is known, the less accurately the other can be measured. The product of the uncertainty in position, and the uncertainty in momentum, is always greater than or equal to h/(2pi). While this is neglegible on the macroscopic level, subatomically, it causes great uncertainty. As a result, the world exists in statistical probabilities.

For example, a electron cannot be found precisely at one location. At any given chance, it has a certain probability of being found anywhere in the universe, with higher probabilities in regions around the atomic nucleus and less likelihood elsewhere. This can be calculated with the wave function.

Quantum Atomic Model

The quantum model of the atom calls for arbitrarily defined orbitals. An orbital is a region around the nucleus where an electron has 90% chance of being found. Electrons have spin, a term used for the intrinsic angular momenta subatomic particles have. They have either a spin of +1/2 or -1/2. Two electrons in the same orbital of the same energy level must have opposite spin (Pauli Exclsion Principle).

Schroedinger's Cat

Since the wave function describes the world in terms of statistical probabilities and not certainties, a cat placed in an isolated system with equal chances of being found alive and dead at anytime *is* equally dead and alive at the same time. When an observer is introduced to the isolated system, the wave function collapses and the cat is either dead or alive.

PRINCIPLES OF QUANTUM MECHANICS

UNCERTAINTY PRINCIPLE

One of the most important concepts of quantum mechanics is the uncertainty principle, which states that there is always some uncertainty when trying to measure the posistion and momentum of a particle. More specifically, the product of the uncertainty in position, D *x*, and the uncertainty in momentum, D*m*u, is always greater than or equal to *h* over 2p, where *h* is planck's constant. When trying to determine the position of an electron, for example, we cannot know *exactly* where it is and where it's going at the same time. The more accurately one is measured, the less accurately the other is known.

To we measure the position of the electron, for example, we can only obtain the information through the collision of the electron with another subatomic particle, such as a photon. After the collision, the momentum of the photon changes—it "bounces" away, and from this change in momentum the position of the electron is calculated . However, the photon also gives the electron a change in momentum, so that now the position and velocity of the electron has changed from what it was. Thus, we can't know precisely the position and velocity simultaneously.

ELECTRON IN THE BOHR ATOM

In the Bohr model, electrons revolved around the nucleus in orbits. According to classical physics, they should radiate energy as they move and ultimately fall to the centre of the atom. Yet this is not so. The Quantum explanation is that electrons can only exist in predefined, discrete energy levels, and nowhere in between. The lowest energy level is not at the centre of the atom, and electrons may go no closer than the lowest level. This explains why they don't collide into the nucleus. To move to another energy level, they must gain or lose *exactly* the energy difference between the levels. This explains spectral lines. Another explanation is that, if electrons crashed into the nucleus, their velovity would be zero and the uncertainty would also be zero. This would be a direct violation of the uncertainty principle, a fundamental property of nature.

SPIN AND QUANTUM NUMBERS

It was discovered experimentally that subatomic particles have intrinsic angular momentum. That is, that angular momentum is not given by some interaction, but a property of the particles. This is a quantum property that classical physics has no explanation for. The closest thing on the macroscopic level is rotation (since angular momentum leads to rotating), so this property is called "spin". Subatomic particles, however, don't really spin; they just have the angular momenta. This property is quantatized since the particles may only have certain discrete values of spin and not any value in between. Electrons have spins of either +1/2 or -1/2.

In the quantum model, electrons in the atom are given four quantum numbers to distinguish them. The first three distinguish the energy level, the shape of the orbital, and in which orbital the electron is located (has 90% chance of being found), and the last one gives the spin. The Pauli Exclusion Principle states that no 2 electrons in the atom can have the same 4 quantum numbers. If they occupy the same energy level and orbital, then they must have opposite spin.

NUCLEUS

The nucleus is a compact and dense region in the centre of the atom that makes up for most of its mass. It is made up of a group of subatomic particles known collectively as nucleons. The two states of nucleons are protons and neutrons.

Although the electron, and not the nucleus, is responsable for chemical reactions, the nucleus still holds much information about the atom. Since nucleons are about 2000 times heavier than electrons, the mass of the nucleus is approximately equal to the mass of the atom. The number of protons in the nucleus determines the type of atom, and nuclear charge is the property by which the elements are arranged in the periodic table.

NUCLEAR ENERGY

A tremendous amount of energy is used to hold the nucleus together, since the repulsion between the positively charged protons is great. The bonding forces that hold the nucleus together are collectively known as the nucleur forces. When heavy element nuclei are split into smaller nuclei, or light element nuclei combined to make one nucleus, tremendous energies are released. The former is known fission, and the latter, fusion, and they are the driving forces behind stars and nuclear weopons.

HISTORY

From Dalton's theory, it was thought that atoms were indivisible particles that made up everything in the world. When Rutherford made his gold foil experiment, however, he found that when high-velocity positive alpha particles bombarded a thing metal foil, while most passed straight through, some were reflected back or deflected. This led him to postulate that most of the atom was empty space, with most of the mass concentrated in one dense region in the centre of the arom. Thus, most of the positive alpha particles passed through the empty space, but some hit the dense core and were reflected back by electric repulsion. He called this the nucleus of the atom.

Subsequently, improvements were made on the Rutherford model. The nucleu was theorized to contain both positive and neutral particles.

APPLICATION

The interactions in the nucleus, the strong and the weak forces, are the strongest forces in nature. Processes such as fission and fusion release much of this energy, and make possible new energy sources. In fact, fusion is the process by which stars generate light and heat, and thus the source from which all life on earth is derived. Other processes, such as radiation, have medical applications.

STRUCTURE OF NUCLEUS

The Dalton atomic model suspected no nucleus. It assumed that atoms were the fundamental building blocks of nature and no particles exist that are smaller than atoms. However, Rutherford's Gold-foil experiment led him to suspect that atoms had nuclei where most of its mass was located, and that the nucleus was positive. Bombarding Nitrogen with alpha particles, he noticed that the disintegration of the nucleus produced a positive particle equal to the mass of a hydrogen atom. He called this particle the proton and postulated that they made up the nucleus.

But a lot of positive charge concentrated in a small region would cause a lot of repulsion forces. So when Chadwick discovered the neutron in 1932, Heisenberg created the proton-neutron model of the nucleus. It also successfully explains isotopes.

THE PROTON

The proton is the nucleon that has a positive unit chage. Its mass is one atomic mass unit (amu, or 1.67×10^{-27} kg, 1,836 times the mass of an electron). It is present in the nucleus of every element. Protons are responsible for the nuclear charge, the positive charge of the nucleus that deflected Rutherford's alpha particles. This charge also corresponds to the atomic number, which determines what kind of element the atom belongs to, and where on the periodic table the element belongs.

THE NEUTRON

THe neutron is an electrically neutral particle slightly heavier than the proton. They are present in numbers greater than or equal to the number of protons in every element except Hydrogen (whose nucleus is a single proton), and variations in the number of neutrons of elements accound for isotopes.

The neutron has a maganetic moment, which suggests that its internal structure is made up of electrically charged components whose net charge is zero. The neutron is actually the fusion of a proton and an electron (since in beta decay a neutron is turned into a proton and an electron) and a neutrino.

ISOTOPES

Isotopes are atoms of the same element that have differenet mass. This is explained by the proton-neutron model of the nucleus. The number of protons account for which element it is and thus how it behaves, but the number of neutrons there are determines which isotope it is, and there can be more neutrons than there are protons.

For example, elemental Hydrogen has one proton (H-1) and a mass of one amu. But an isotope of Hydrogen, deuterium, has a proton and a neutron, and a mass of two amu. As another example, Carbon-12, Carbon-13 and Carbon-14 all have 6 protons (this determines that it is Carbon atom) but have 6, 7 and 8 neutrons, respectively.

Sometimes there are too many neutrons in the nucleus. When this happens, the neutron goes under beta decay, a type of radioactive decay.

The neutron splits into a proton, an electron (beta particle) and a neutrino. The proton stays in the nucleus, but the beta particle and the neutrino(an electrically neutral, light or massless particle) are emitted, and energy is released, carried by the electron (when the electron doesn't carry all the energy, the rest is carried by the neutrino).

CLASSIFICATION

Nucleons are generally knoen as a baryon. Baryons are elementary particles consisting of three quarks and no antiquarks. They obey the Fermi-Dirac statistics, which encompass all elementary particles that obey the Pauli Exclusion Principle. The proton is the lightest baryon and is made of two up-quarks and one down-quark. The neutron is made of two down-quarks and one up-quark.

Nucleon's antiparticles are the antiproton and the antineutron, which have the same mass as the corresponding nucleons but opposite electric charge and magnetic moment.

TYPES OF RADIOACTIVITY

The study of radioactive material started with Becqueral in 1896, when he discovered that some of his films became unuseable, as if they had been exposed to light, even though they were kept in a sealed opaque container. Further investigation revealed a substance that gave a green glow in the dark. This was uranium, and Becqueral discovered the X-ray.

Marie Curie then studied radioactive material, discovering several radioactive substances. This research led to a better understanding of radiation, and peopls started bombarding elements with particles emitted from radiation. Rutherford's experimentations, for example, led him to his atomic model. These bombardments also led to the discovery of many particles, such as the proton. The nature of radiation was also better understood as the identity of the radiated particles came to be known, and theories of radioactive decay were devised.

These experiments did not reveal a complete picture of radiation, until Fermi's systematic bombardment of all elements. The results led him to envision a chain reaction, where the bombardment of an atom led to its splitting and emitting more particles that bombarded the other atoms. This was carried out in 1942, and led to the development of the A-bomb shortly after.

ALPHA PARTICLES

Alpha radiation (or alpha rays) was distinguished and named by E. R. Rutherford in 1909, who found by measuring the charge and mass of alpha particles that they are the nuclei of ordinary helium atoms. Alpha particles consist of two protons and two neutrons.

BETA PARTICLES

Beta radiation (or beta rays) was identified and named by E. Rutherford too. He found that beta rays consists of high-speed electrons. Unlike alpha and gamma particles, whose energy can be explained as the difference of the energies of the radioactive nucleus before and after emission, beta particles emerge with a variable energy.

This apparent violation of the law of conservation of energy led to the hypothesis that a second undetected particle, the neutrino, is emitted along with the electron and shares the total available energy. In some forms of induced, or artificial, radioactivity, the electron's antiparticle, the positron, is emitted from the excited nucleus; the positron in this case is also called a beta particle and denoted by b^+ (the ordinary beta particle is b^-).

GAMMA RADIATION

Gamma radiation is high-energy photons emitted as one of the three types of radiation resulting from natural radioactivity. It is the most energetic form of electromagnetic radiation, with a very short wavelength (high frequency). Wavelengths of the longest gamma radiation are less than 10^{-10} m, with frequencies greater than 10^{18} hertz (cycles per sec). Gamma rays are essentially very energetic X rays; the distinction between the two is not based on their intrinsic nature but rather on their origins.

X rays are emitted during atomic processes involving energetic electrons. Gamma radiation is emitted by excited nuclei or other processes involving subatomic particles; it often accompanies alpha or beta radiation, as a nucleus emitting those particles may be left in an excited (higher-energy) state. The applications of gamma radiation are much the same as those of X rays, both in medicine and in industry. In medicine, gamma ray sources are used for cancer treatment and for diagnostic purposes.

Some gamma-emitting radioisotopes are also used as tracers. In industry, principal applications include inspection of castings and welds.

Data from artificial satellites and high-altitude balloons have indicated that a flux of gamma radiation is reaching the earth from outer space, thus opening up the field of research known as gamma-ray astronomy.

DECAY

ALPHA DECAY

There are three types of radioactive decay, named by the type of particle they produced. All radioactive decay is the result of an unstable nucleus and release energy. The first is the alpha decay, which emits an alpha particle, a positively charged and heavy particle that does not penetrate too well. It was shown that the alpha particle is actually a Helium nucleus, consisting of two protons and two neutrons. When the unstable nucleus emits the alpha particle, the atomic number of the atom decreases by two and the atomic mass number decreases by four.

BETA DECAY

The beta particle is an electron. When beta decay occurs, a neutron in the nucleus converts into a proton and an electron. The electron is emitted and takes the released energy with it. When it does not carry enough enery, however, the remaining is carried by a small, mass less (or very light) neutral particle called the neutrino. The beta particle is therefore negative and more penetrating than the alpha particle.

GAMMA DECAY

The Gamma radiation is unaffected by magnetic field as it travels; therefore it is electrically neutral. It is actually a type of electromagnetic radiation, and not a particle of significant mass.

Gamma rays is one of the most energetic types of electromagnetic radiation, because its wavelength is very short and its frequency very high, (they are therefore invisible to the eye). Gamma radiation often accompanies alpha decay or beta decay, since the nucleus emitting those particles may still be excited, and the gamma ray is the result of the extra energy. Gamma rays are the most penetrating type of radiation.

NATURAL GROWTH

Radioactive decay obeys the rules of natural (exponential) growth or decay. That is, the rate of decay is directly proportional to the amount of undecayed substance present. (Chemically, the rate of reaction is directly proportional to the amount of unreacted reactant.) Thus, the first derivative of the substance, say, Uranuim (U) with respect to time is proportional to the amount of undecayed U present

$$d[U]/dt = k[U].$$

Solving this gives $d[U]/[U] = kdt$, or

$\ln[U] + C = kt$, $[U] = a^*e(kt)$,

where a is calculated to be the amounf or Uranium at the start of the decay.

A proporty of the exponential graph is that starting at any arbitrarily chosen time, the time it takes for the amount of undecayed Uranium to decrease by half (or any fraction) is the same.

This period of time, the time it take for half the sample to undergo radioactive decay, is called the Half-life of that element. After one half-life, the amount of unreacted sample left is alway one half the amount of unreacted sample before that period of time.

Thus, if the half life of a particular sample is 10 years, then after 20 years one-quarter of undecayed sample will be left.

INDUCED RADIOACTIVITY

The induced radioactivity was discovered by French physicists Iren Joliot-Curie (1897-1956) and Frederic Joliot-Curie (1900-1960). The first synthetic nuclear reaction was realized by Rutherford. It is possible to register the charged particles which are taking off at a radioactive decay. It allows to use atoms of the synthetic radioactive isotopes.

NUCLEAR REACTIONS

Nuclear reactions. There are 2 stages of nuclear reactions. The first stage of nuclear reaction consists in entrapment by a kern of a particle. For a small time interval the energy is uniformly proportioned between all particles of a compound nucleus.

The second stage of nuclear response (start of a particle from a compound nucleus) occurs as a result of casual diversions from a uniform distribution of energy. There is a time between entrapment of particles and their start from a kern, which is significant more then nuclear time $\tau_n = 10^{-22}$ s. All process of nuclear reaction is possible to figure by the formula:

$$_{Z1}X^{A1} + a \rightarrow _{Z2}Y^{A2} \rightarrow _{Z3}C^{A3} + b$$

$_{Z1}X^{A1}$ -inital kern *a, b* particles

$_{Z2}Y^{A2}$ -compound nucleus; $_{Z3}C^{A3}$ -terminating kem

DIVISION OF KERNS

The heavy kerns are labile. The kern, which was exited as a result of a neutron capture, can be divided into identical parts. The nuclear fission on two debrises is accompanied by eduction of huge energy.

The debrises, which are formed at division, are radioactive and can emit neutrons. As the result the exuberant number of neutrons is formed, that gives in radio-activity of yields of decay.

REACTION OF DIVISION

CHAIN REACTION OF DIVISION

The division of kerns of uranium occurs at interaction to prompt or sluggish neutrons, but the irradiation by sluggish neutrons is more effective. The practical interest is represented by reactions, at which the major energy is oozed and there are neutrons, which give in magnification of number of the acts of division. For realization of chain reaction the maintenance of fixed number of neutrons is necessary.

For an estimation of number of neutrons is inlet coefficient of manifolding of neutrons k. It determines the ratio of number of neutrons, which arise at reaction, to number of neutrons, which participated in reaction. If $k >= 1$, the reaction will be explicated.

The minimum sizes of a fissile region, at which the realization of a chain reaction is possible, term as the critical sizes. The minimum quantity of a fissionable material necessary to sustain a nuclear chain reaction is called the critical mass.

CONTROLLABLE REACTION OF DIVISION

The controllable reactions occurs in a spacial nuclear reactors. Reactors can be used for research or for power production. A research reactor is designed to produce various beams of radiation for experimental application; the heat produced is a waste product and is dissipated as efficiently as possible.

In a power reactor the heat produced is of primary importance for use in driving conventional heat engines; the beams of radiation are controlled by shielding.

A-BOMB

Practical fissionable nuclei for atomic bombs are the isotopes uranium-235 and plutonium-239, which are capable of undergoing chain reaction. If the mass of the fissionable material exceeds the critical mass (a few pounds), the chain reaction multiplies rapidly into an uncontrollable release of energy. An atomic bomb is detonated by bringing together very rapidly (e.g., by means of a chemical explosive) two sub critical masses of fissionable material, the combined mass exceeding the critical mass.

An atomic bomb explosion produces, in addition to the shock wave accompanying any explosion, intense neutron and gamma radiation, both of which are very damaging to living tissue. The neighborhood of the explosion becomes contaminated with radioactive fission products. Some radioactive products are borne into the upper atmosphere as dust or gas and may subsequently be deposited partially decayed as radioactive fallout far from the site of the explosion.

ELEMENTARY PARTICLES

All subatomic particles, such as nucleons or electrons, and the particles they release when undergoing processes such as radioactive decay, are organized into classes.

All particles that obey the Pauli Exclusion Principle (no two particles can have the same four quantum numbers in a system, such as an atom) are called fermions because the Fermi-Dirac statistics apply to them. All particles that don't obey the principle are called bosons because the Bose-Einstein statistics apply to them.

FERMIONS

Fermions have half-integral spin (intrinsic angular momentum). They are deemed anitsocial particles because they obey the exculsion principle, meaning that no two fermion with the same spin can be found in a system. The antiparticles for fermions have the same mass but opposite charge and magnetic moment as the corresponding particle.

BARYONS

The first type of fermion is the baryon. They are those particles that experience the strong nuclear force. They consist of three quarks (quark triplet). The most notable baryons are the nucleons, the two states of which are protons and neutrons, the particles that make up the nucleus. Other, heavier baryons, called hyperons also exist, but they are unstable, and usually decays to a form of nucleon(heavier baryons include the sigma, the lamda and the delta particles). Since all baryons decay to the nucleon form, the nucleons are known as the baryon ground state.

All baryons obey the law of baryon family numbers (the family number is +1 for particles such as protons and -1 for antiparticles such as antiprotons), which states that the sum of the family number of all reactants in an interaction is equal to the sum of the family number of all products. This means that, for every particle created or destroyed, a corresponding antiparticle is also created or destroyed. For example, a proton colliding with an antiproton will result in the mutual destruction of both particles, or, a proton and an antiproton can be created simultaneously.

LEPTONS

Leptons are fermions that do not experience the strong nuclear force. They are only weakly interactive, and can be created by nuclear decay. Leptons include such particles as electrons, muons, neutrinos and the tau particle. So, for example, an electron and a neutrino can be created by the decay of an unstable neutron to a proton. As fermions they obey the exclusion principle, so no two electrons can occupy the same energy level and orbital of an atom (a system) and have the same spin. Leptons are the lightest fermions.

There are three separate laws concerning the conservation of lepton family numbers (the family numbers for leptons, like for baryons, is +1 for particles and -1 for antiparticles), for the electrons, muons and taus.

BOSONS

Bosons are particles that, unlike the fermions, do not obey the exclusion principle. Generally, fermions are particles that compose atomic structure but bosons carry out the interaction (forces) between fermions. Bosons have integral spin and are deemed social because many bosons can by placed in the same system in the same energy state at once.

MESONS

Mesons are a class of bosons whose mass falls between the lighter lepton fermions and the heavier baryon fermions. They consist of two quarks (quark pair), and include such particles as the p meson (pion) and the *K* meson (kaon). It was previously thought that the muon was a type of meson (m meson) because of its large mass (about 200 times that of the electron), but the behaviour of the muon doesn't fit into the description of mesons, so it was reclassified.

The pion is accountable for carrying out the strong nuclear force, which is the force that holdes the nucleus together, despite trememndous electric repulsion.

OTHER BOSONS

Other bosons include the W and Z particles are responsible for carrying out the weak nuclear force and wjocj are about 200 heavier than the proton, and the photon, the particle form of light. The photon is its own antiparticple, as is the neutral pion.

Chapter 3

Radioactivity

Radiation is all around us. It comes from the Earth and from outer space. Many forms of radiation are invisible — we can't feel it, see it, taste it, or smell it. Yet, it can be detected and measured when present. We measure ionizing radiation in units called millirems. But what is radiation? Radioactive materials are composed of atoms that are unstable. An unstable atom gives off its excess energy until it becomes stable. The energy emitted is radiation. We can classify radiation as being either natural or man-made.

The Earth is surrounded by radiation. Every day, for example, we are exposed to radon, a radioactive gas from uranium found in soil dispersed in the air; from radioactive potassium in our food and water; from uranium, radium, and thorium in the Earth's crust; and from cosmic rays and the sun.

These types of radiation are called natural or background radiation. We are exposed to an average of 300 millirems of natural radiation each year (a millirem is a unit of measure for exposure to radiation). This amounts to natural radiation accounting for nearly 85 percent of our total annual exposure. Where does the remaining 15 percent come from? Man-made sources.

Man-made radiation sources that people can be exposed to include tobacco, television, medical x-rays, smoke detectors, lantern mantles, nuclear medicine, and building materials.Generally, when we think of exposure to radiation, we need to look at radioactive atoms produced in nuclear reactors and described as being unstable. They are unstable because they undergo a disintegrating process called decaying. During this process, unstable atoms becomes stable, throwing off (emitting) radiation in the form of rays and/or particles.How fast a radioactive atom decays into a stable atom depends on the atom itself. For example, the range in the rate of decay among isotopes goes from fractions of seconds to several billion years (e.g., uranium).Let's take a look at uranium-238 to the decay chain.

U_{238} →
Thorium$_{230}$ →
Radium$_{226}$ →
Radon$_{218}$ →
Bismuth$_{214}$ →
Lead$_{206}$

As U-238 decays it changes into thorium-230, which changes into radium-226, which changes into radon-218, which changes into bismuth-214, and finally into lead-206 (a stable element).

One peculiar thing about radioactive atoms is that no one knows exactly when the element will decay and give off radiation. There is, however, a pattern relating to how long it takes for an isotope to lose half of its radioactivity. The pattern is called half-life. If an atom, for example, has a half-life of 10 years, half of its atoms will decay in 10 years. Then in another 10 years half of that amount will decay and so on. While there are several different forms of radiation, we're going to concentrate on just three that result from the decay of radioactive isotopes: alpha, beta, and gamma.

Beta particles are high energy electrons. Both alpha and beta particles are emitted from unstable isotopes. The alpha particle, consisting of two protons and two neutrons, is relatively large compared to beta particles. Gamma rays have no mass.

Because of its size and electrical charge (+2), the alpha particle has a relatively slow speed and low penetrating distance (one to two inches in air). Alpha particles are easily stopped by a thin sheet of paper or the body's outer layer of skin. Since they do not penetrate the outer (dead) layer of skin, they present little or no hazard when they are external to the body.

However, alpha particles are considered internal hazards, because when they come into contact with live tissue they cause a large number of ionizations to occur in small areas, thus causing damage to tissues and cells.

Beta radiation, while faster and lighter than alpha radiation, can travel through about 10 feet of air and penetrate very thin layers of materials such as aluminum foil. However, while clothing will stop most beta particles, they can penetrate the live layers of skin tissue.

Therefore, beta radiation is considered to be both an internal and external (to skin only) hazard. Thin layers of metals and plastics can be used to shield individuals from beta radiation.

Gamma radiation, high energy light, is a little different. It is a type of electromagnetic wave, just like radio waves, light waves, and x-rays. Gamma radiation is a very strong type of electromagnetic wave. It is has no weight and travels at the speed of light. This is much faster than alpha and beta radiation.

Because of their penetrating capability, gamma rays are considered both internal and external hazards. Thick walls of cement, lead, or steel are needed to stop it.

Alpha, beta, and gamma radiations are also known as ionizing radiation. Ionizing radiation is especially harmful because it can change the chemical makeup of many things, including the delicate chemistry of the human body and other living organisms. For this reason it is a good idea to avoid unnecessary exposure to all ionizing radiation.

The problem is simply this: large amounts of radiation — far above the levels encountered in daily life — can produce cancer and genetic defects in living organisms. Radiation causes damage and alters the body's normal cells and normal cell function. This breakdown in normal cell function may result in an uncontrolled growth of cells, hence the potential for malignant/cancerous tumors.

Whether the source of radiation is natural or man-made, whether it is a small dose of radiation or a large dose, there will be some biological effects. A diagram can show the biological effect of ionizing radiation. Radiation causes ionizations of atoms that will affect molecules that may affect cells that may effect tissues that may affect organs that may affect the whole body.

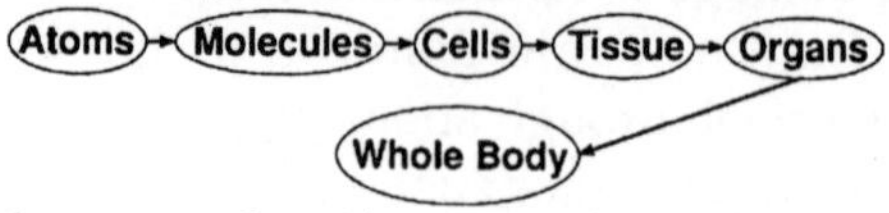

Although we tend to think of biological effects in terms of the effect of radiation on living cells, in actuality, ionizing radiation, by definition, interacts only with atoms by a process called ionization. Thus, all biological damage effects begin when radiation interacts with atoms forming the cells in the human body. As a result, radiation effects on humans proceed from the lowest to the highest level as noted on the board.

Although scientists have only known about radiation since the 1890s, they have developed a wide variety of uses for this natural force. Today, to benefit mankind, radiation is used in science, medicine, and industry, as well as for generating electricity. Radiation has useful applications in such areas as agriculture, medicine, space exploration, architect/engineering, industry/manufacturing, government, geology (including mining), ecology, and education.

Radiation is used by doctors to diagnose illness and helps archaeologists find the age of ancient artifacts. Electricity produced by nuclear fission — splitting the atom — is one of its greatest uses. A reliable source of electricity is needed to give us light, help to groom and feed us, and to keep our homes and businesses running. Let me give you some specific examples of how the radiation has been used to -

- Diagnose and treat illnesses.
- Kill bacteria and preserve food without chemicals and refrigeration.
- Process sludge for fertilizer and soil conditioner.
- Locate underground natural resources and tell a dry hole from a gusher.

- Make smoke detectors, nonstick fry pans, and ice cream.
- Grow stronger crops.
- Power satellites and provide future electrical needs for space laboratories with people on board.
- Design instruments, techniques, and equipment; measure air pollution.
- Prove the age of works of art and assist in determining their authenticity.

RADIATION IN SCIENCE

Radiation is used in science in many ways. Just as doctors can label substances inside people's bodies, scientists can label substances that pass through plants, animals, or our world. This allows us to study such things as the paths that different types of air and water pollution take through the environment.

It has also helped us learn more about a wide variety of things, such as what types of soil different plants need to grow, the size of newly discovered oil fields, and the track of ocean currents.

Scientists use radioactive substances to find the age of ancient objects by a process called carbon dating. For example, in the upper levels of our atmosphere, cosmic rays hit nitrogen atoms and form a naturally radioactive isotope called carbon-14. Carbon is found in all living things, and a small percentage of this carbon is carbon-14. When a plant or animal dies, it no longer takes in new carbon and the carbon-14 it contains begins the process of radioactive decay.However, new isotopes of carbon-14 continue to be formed in our atmosphere, and after a few years the percent of radioactivity in an old object is less than it is in a newer one. By measuring this difference, scientists are able to determine how old certain objects are. The measuring process is called carbon dating.

RADIATION USED TO SOLVE CRIMES

How is radiation used to solve crimes? I am sure that you already know that detectives often search the scene of a crime for traces of paint, glass, hair, gunpowder, or blood. But you may not know that after such evidence is collected, it is often exposed to radiation and then analyzed to find out its exact makeup.

If material is exposed to streams of neutrons, some of the neutrons can be absorbed into the nucleus of the exposed material. This makes these materials slightly radioactive because they are unstable and decay with time. Scientists are then able to read the exact chemical signatures of these substances. This laboratory process, called activation analysis, is precise enough to tell if a single hair found at a crime scene came from a certain person.

Activation analysis is also used to find out the chemical makeup of materials when scientists only have small samples, as well as to prove that older works of art are not made of modern materials.

RADIATION IN INDUSTRY

We could talk all day about the many and varied uses of radiation in industry and not complete the list. To make a long story short, we'll just concentrate on a few. Exposure to some types of radiation (for example, x-rays) can kill germs without harming the items that are being disinfected or making them radioactive. For example, when treated with radiation, foods take much longer to spoil, and medical equipment such as bandages, hypodermic syringes, and surgical instruments don't have to be exposed to toxic chemicals or extreme heat to be sterilized. Although we now use chlorine, a toxic and difficult-to-handle chemical, we may use radiation in the future to disinfect our drinking water and even kill all the germs in our sewage. Ultraviolet light already is used to disinfect drinking water in some homes.

The agricultural industry makes use of radiation to improve food production. Plant seeds, for example, have been exposed to radiation to bring about new and better types of plants. Besides making plants stronger, radiation can also be used to control insect populations, thereby decreasing the use of pesticides. Engineers use radioactive substances to measure the thickness of materials and an x-ray process called radiography to find hard to detect defects in many types of metals and machines. Radiography is also used to check such things as the flow of oil in sealed engines and the rate and way various materials wear out. And we've already talked about the use of the radioactive element uranium, which is used as a fuel to make electricity for our cities, farms, towns, factories, etc.

In outer space, radioactive materials are also used to power space craft. Such materials have also been used to supply electricity to satellites sent on missions to the outermost regions of our solar system.

Radiation has been used to help clean up toxic pollutants, such as exhaust gases from coal-fired power stations and industry. Sulphur dioxides and nitrogen oxides, for example, can be removed by electron beam radiation.

As you can see, radiation and radioactive materials have played and will continue to play a very significant role in our lives. Let's sum up this discussion with a walk through the life of a typical family for one day and learn about some of the uses of radiation.

Dad gets up in the morning and puts on a clean shirt. His polyester-cotton blend shirt is made from chemically treated fabric that has been irradiated (treated with radiation) before being exposed to a soil-releasing agent. The radiation makes the chemicals bind to the fabric, keeping his shirt fresh and pressed all day. The shirt is not radioactive.

In the kitchen, Jenny is frying an egg. That nonstick pan she is using has been treated with gamma rays, and the thickness of the eggshell was measured by a gauge containing radioactive material before going into the egg carton. Thin, breakable eggs were screened out. The turkey Mom is taking out of the refrigerator for tonight's dinner was covered with irradiated polyethylene shrink wrap. Once polyethylene has been irradiated, it can be heated above its usual melting point and wrapped around the turkey to provide an airtight cover.

As Dad drives to work, he passes reflective signs that have been treated with radioactive tritium and phosphorescent paint. During lunch, brother Bob has some ice cream. The amount of air whipped into that ice cream was measured by a radioisotopic gauge. After you and your family return home this evening, some of you may have soda and others may sit and relax. Nuclear science is at work here: The soda bottle was carefully filled — a radiation detector prevented spillover. And your family is safe at home because the ionizing smoke detector, using a tiny bit of americium-241, will keep watch over you while you sleep.

Increasingly, our country has become a nation of electricity users. We depend on an abundant affordable supply of energy to power the many machines we use in our complex society. Can you imagine what it would be like not having electricity in your home? About one-third of our energy resources are used to produce electricity.

Electricity can be produced in many ways — most of which you already know about. Today, we're going to talk about one of those ways — nuclear fission. In America, nuclear energy plants are the second largest source of electricity after coal — producing approximately 21 percent of our electricity. There is something I want all of you to be aware of: The purpose of a nuclear power plant is to produce electricity.

While nuclear power plants have many similarities to other types of electricity generating plants, there are some significant differences.

With the exception of solar, wind, and hydroelectric plants, all others including nuclear convert water to steam that spins the propeller-like blades of a turbine that spins the shaft of a generator. Inside the generator coils of wire and magnetic fields interact to create electricity.

The energy needed to boil water into steam is produced in one of two ways: by burning coal, oil, or gas (fossil fuels) in a furnace or by splitting certain atoms of uranium in a nuclear energy plant. Nothing is burned or exploded in a nuclear energy plant. Rather, the uranium fuel generates heat through a process called fission.

Uranium is an element that can be found in the crust of the Earth. This element, quite abundant in many areas of the world, is naturally radioactive. Certain isotopes of uranium can be used as fuel in a nuclear power plant. The uranium is formed into ceramic pellets about the size of the end of your finger. These pellets are inserted into long, vertical tubes (fuel rods) within the reactor. The reactor is the heart of the nuclear power plant. Basically, it is a machine that heats water.

A reactor has four main parts: the uranium fuel assemblies, the control rods, the coolant/moderator, and the pressure vessel. The fuel assemblies, control rods, and coolant/moderator make up what is known as the reactor core. The core is surrounded by the pressure vessel.

We also have to understand that uranium cannot just be thrown into a reactor the way we shovel coal into a furnace. The fuel rods, containing the uranium, are carefully bound together into fuel assemblies, each of which contains about 240 rods. The assemblies hold the rods apart so that when they are submerged into the reactor core, water can flow between them.

When the uranium atom splits, it releases energy and two or more neutrons from its nucleus. These neutrons can then hit the nuclei of other uranium atoms causing them to fission. The sequence of one fission triggering others, and those triggering still more, is called a chain reaction. When the atoms split, they release energy in the form of heat. The heat is transferred from the reactor core to the steam generator. Here, it's converted to high pressure steam which turns the turbine in the electric generator.

This next transparency shows the basic design of the nuclear fuel steam plant and its various components. In the past few minutes I've introduced several new terms. Let me explain what each means in the energy generating process. First, control rods. What are they? How are they used?

The control rods slide up and down in between the fuel assemblies in the reactor core. They control or regulate the speed of the nuclear reaction by absorbing neutrons. Here's how it works: When the control rods absorb neutrons, fewer neutrons hit the uranium atoms thus slowing down the chain reaction. On the other hand, when the core temperature goes down, the control rods are slowly lifted out of the core, and fewer neutrons are absorbed. Therefore, more neutrons are available to cause fission. This releases more heat energy.

Just as there are different types of houses and cars, there are different types of nuclear power plants that generate electricity. The two basic types being used are the boiling water reactor (BWR) and the pressurized water reactor (PWR). These power plants are often referred to as light water reactors.

BOILING WATER REACTOR (BWR)

The boiling water reactor operates in essentially the same way as a fossil fuel generating plant. Neither of these types of power plants have a steam generator. Instead, water in the BWR boils inside the pressure vessel and the steam water mixture is produced when very pure water (reactor coolant) moves upward through the core absorbing heat. The water boils and produces steam. When the steam rises to the top of the pressure vessel, water droplets are removed, the steam is sent to the turbine generator to turn the turbine.

PRESSURIZED WATER REACTOR (PWR)

The pressurized water reactor differs from the BWR in that the steam to run the turbine is produced in a steam generator. Water boils at 212°F or 100°C. If a lid is tightly placed over a pot of boiling water (a pressure cooker), the pressure inside the pot will increase because the steam cannot escape. As the pressure increases, so does the temperature of the water in the pot. In the PWR plant, a pressurizer unit keeps the water that is flowing through the reactor vessel under very high pressure to prevent it from boiling. The hot water then flows into the steam generator where it is converted to steam. The steam passes through the turbine which produces electricity.

Chapter 4

Atomic Structure

DISCOVERY OF α, β AND y PARTICLES

The discovery of x-rays by William Conrad Roentgen in November of 1895 excited the imagination of a generation of scientists who rushed to study this phenomenon. Within a few months, Henri Becquerel found that both uranium metal and salts of this element gave off a different form of radiation, which could also pass through solids. By 1898, Marie Curie found that compounds of thorium were also "radioactive." After pain-staking effort she eventually isolated two more radioactive elements polonium and radiumfrom ores that contained uranium.

In 1899 Ernest Rutherford found that there were at least two different forms of radioactivity when he studied the absorption of radioactivity by thin sheets of metal foil. One, which he called alpha (α) particles, were absorbed by metal foil that was a few hundredths of a centimeter thick. The other, beta (β) particles, could pass through 100 times as much metal foil before they became absorbed. Shortly thereafter, a third form of radiation, gamma (γ) rays, was discovered that could penetrate as much as several centimeters of lead.

The direction in which -particles were deflected by an electric field suggested that they were positively charged. The magnitude of this deflection suggested that they had the same charge-to-mass ratio as an He^{2+} ion. To test the equivalence between -particles and He^{2+} ions, Rutherford built an apparatus that allowed -particles to pass through a very thin glass wall into an evacuated flask that contained a pair of metal electrodes. After a few days, he connected these electrodes to a battery and noted that the gas in the flask did indeed give off the characteristic emission spectrum of helium.

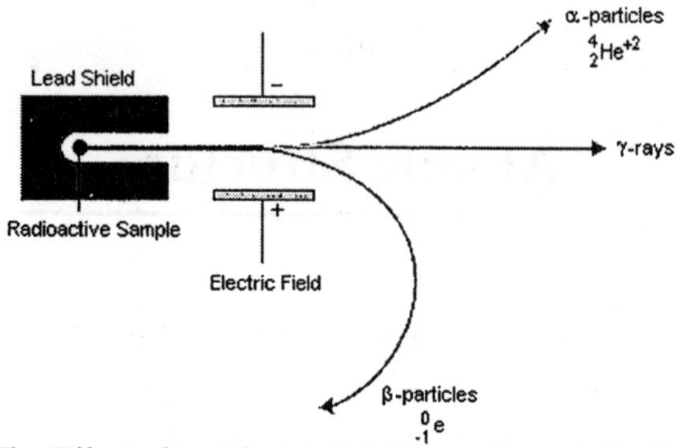

Fig. The Effect of an Electric F ield on -, b, and -Radiation.

Experiments with electric and magnetic fields demonstrated that β particles were negatively charged. Furthermore, they had the same charge-to-mass ratio as an electron. To date, no detectable difference has been found between β -particles and electrons. The only reason to retain the name " β -particle" is to emphasize the fact that these particles are ejected from the nucleus of an atom when it undergoes radioactive decay.

The fact that -rays are not deflected by either electric or magnetic fields suggests that these rays don't carry an electric charge. Since they travel at the speed of light, they are classified as a form of electromagnetic radiation that carries even more energy than x-rays.

GROWTH AND DECAY CURVES

At the turn of the century, when radioactivity was discovered, atoms were assumed to be indestructible. Ernest Rutherford and Frederick Soddy, however, found that radioactive substances became less active with time. More importantly, they noticed that radioactivity was always accompanied by the formation of atoms of a different element. By 1903, they concluded that radioactivity was accompanied by a change in the structure of the atom. They therefore assumed that radiation was emitted when an element decayed into a different kind of atom.By 1910, 40 radioactive elements had been isolated that were associated with the process by which uranium metal decayed to lead. This created a problem, however, because there was space for only 11 elements between lead and uranium. In 1913, Kasimir Fajans and Fredick Soddy proposed an explanation for these results based on the following rules.

- α -particles are emitted when an element is formed that belongs two spaces to the left in the periodic table. Uranium (Z = 92), for example, emits an -particle when it decays to form thorium (Z = 90).
- β -particles are emitted when an element is formed that belongs one space to the right in the periodic table. Actinium (Z = 89) emits a β -particle when it decays to form thorium (Z = 90).
- Radioactive elements that fall in the same place in the periodic table are different forms of the same element. The radioactive thorium produced by the -particle decay of uranium is a different form of the element than the radioactive thorium obtained by the β -particle decay of actinium.

Soddy proposed the name isotope to describe different radioactive atoms that occupy the same position in the periodic table. J. J. Thomson and Francis Aston then used a mass spectrometer to show that isotopes are atoms of the same element that have different atomic masses.

THE STRUCTURE OF THE ATOM

The discovery of the electron in 1897 by J. J. Thomson suggested that there was an internal structure to the "indivisible" building blocks of matter known as atoms. This raised an obvious question: How many electrons does an atom contain? By studying the scattering of light, x-rays, and -particles, Thomson concluded that the number of electrons in an atom was between 0.2 and 2 times the weight of the atom.

In 1911, Rutherford concluded that the scattering of -particles by extremely thin pieces of metal foil could be explained by assuming that all of the positive charge and most of the mass of the atom were concentrated in an infinitesimally small fraction of the total volume of the atom, for which he proposed the name nucleus. Rutherford's data also suggested that the nucleus of a gold atom carries a positive charge that is about 80 times the charge on an electron.

The discovery of the neutron in 1932 explained the discrepancy between the charge on the nucleus and the mass of an atom.

A neutral gold atom that has a mass of 197 amu consists of a nucleus that contains 79 protons and 118 neutrons surrounded by 79 electrons. By convention, this information is specified by the following symbol, which describes the only naturally occurring isotope of gold.

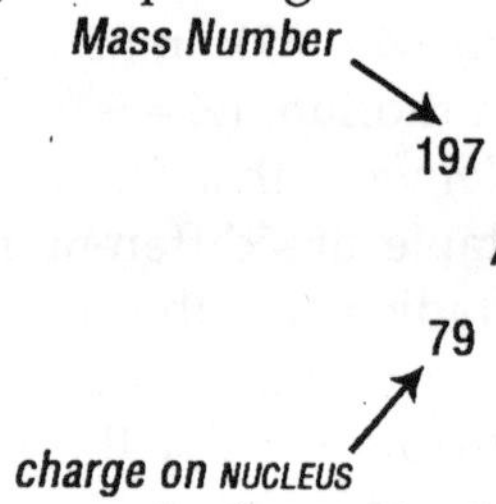

This convention can also be applied to subatomic particles. The only difference is the use of lowercase letters to identify the particle.

$$\begin{matrix} 0 & 1 & 1 \\ -1^{e} & 1^{p} & 0^{n} \end{matrix}$$

electron proton neutron

Because anyone with access to a periodic table can find the atomic number of an element, a shorthand notation is often used that reports only the mass number of the atom and the symbol of the element. The shorthand notation for the naturally occurring isotope of gold is ^{197}Au.

RADIOACTIVE DECAY

α, β and γ Radiation

Early studies of radioactivity indicated that three different kinds of radiation were emitted, symbolized by the first three letters of the Greek alphabet , β and . With time, it became apparent that this classification scheme was much too simple. The emission of a negatively charged β particle, for example, is only one example of a family of radioactive transformations known as β -decay. A fourth category, known as spontaneous fission, also had to be added to describe the process by which certain radioactive nuclides decompose into fragments of different weight.

ALPHA DECAY

Alpha decay is usually restricted to the heavier elements in the periodic table. (Only a handful of nuclides with atomic numbers less than 83 emit an -particle). The product of -decay is easy to predict if we assume that both mass and charge are conserved in nuclear reactions. Alpha decay of the ^{238}U "parent" nuclide, for example, produces ^{234}Th as the "daughter" nuclide.

$$^{238}_{92}U \rightarrow {}^{234}_{90}Th + {}^{4}_{2}He$$

The sum of the mass numbers of the products (234 + 4) is equal to the mass number of the parent nuclide (238), and the sum of the charges on the products (90 + 2) is equal to the charge on the parent nuclide.

BETA DECAY

There are three different modes of beta decay:

- Electron (β^-) emission
- Electron capture
- Positron (β^+) emission

Electron (β^-) emission is literally the process in which an electron is ejected or emitted from the nucleus. When this happens, the charge on the nucleus increases by one. Electron (β^-) emitters are found throughout the periodic table, from the lightest elements (^3H) to the heaviest (^{255}Es). The product of β $^-$-emission can be predicted by assuming that both mass number and charge are conserved in nuclear reactions. If ^{40}K is a β^--emitter, for example, the product of this reaction must be ^{40}Ca.

$$^{40}_{19}K \rightarrow {}^{40}_{20}Ca + {}^{0}_{-1}e$$

Once again the sum of the mass numbers of the products is equal to the mass number of the parent nuclide and the sum of the charge on the products is equal to the charge on the parent nuclide.

Nuclei can also decay by capturing one of the electrons that surround the nucleus. Electron capture leads to a decrease of one in the charge on the nucleus. The energy given off in this reaction is carried by an x-ray photon, which is represented by the symbol hv, where h is Planck's constant and v is the frequency of the x-ray. The product of this reaction can be predicted, once again, by assuming that mass and charge are conserved.

$$^{40}_{19}K + {}^{0}_{-1} \rightarrow {}^{40}_{18}Ar + hv \qquad \text{(Electron capture)}$$

The electron captured by the nucleus in this reaction is usually a 1s electron because electrons in this orbital are the closest to the nucleus.

A third form of beta decay is called positron (β^+) emission. The positron is the antimatter equivalent of an electron. It has the same mass as an electron, but the opposite charge. Positron (β^+) decay produces a daughter nuclide with one less positive charge on the nucleus than the parent.

$$^{40}_{19}K + {}^{40}_{18}Ar + {}^{0}_{+1}e \qquad \text{(Position } (\beta^+) \text{ emission)}$$

Positrons have a very short life-time. They rapidly lose their kinetic energy as they pass through matter. As soon as they come to rest, they combine with an electron to form two -ray photons in a matter-antimatter annihilation reaction.

$$^{0}_{+1}e + {}^{0}_{-1}e \rightarrow 2\gamma$$

Thus, although it is theoretically possible to observe a fourth mode of beta decay corresponding to the capture of a positron, this reaction does not occur in nature.

Note that in all three forms of β -decay for the ^{40}K nuclide the mass number of the parent and daughter nuclides are the same for electron emission, electron capture, and position emission. All three forms of β -decay therefore interconvert isobars.

GAMMA EMISSION

The daughter nuclides produced by -decay or β -decay are often obtained in an excited state. The excess energy associated with this excited state is released when the nucleus emits a photon in the -ray portion of the electromagnetic spectrum.

Most of the time, the -ray is emitted within 10^{-12} seconds after the -particle or β -particle. In some cases, gamma decay is delayed, and a short-lived, or metastable, nuclide is formed, which is identified by a small letter m written after the mass number. ^{60m}Co, for example, is produced by the electron emission of ^{60}Fe.

$$^{60}_{26}Fe \rightarrow ^{60}_{27}Co + ^{0}_{-1}e$$

The metastable ^{60m}Co nuclide has a half-life of 10.5 minutes. Since electromagnetic radiation carries neither charge nor mass, the product of -ray emission by ^{60m}Co is ^{60}Co.

$$^{60}_{26}Co \rightarrow ^{60}_{27}Co + \gamma \qquad (\gamma - \text{ray emission})$$

SPONTANEOUS FISSION

Nuclides with atomic numbers of 90 or more undergo a form of radioactive decay known as spontaneous fission in which the parent nucleus splits into a pair of smaller nuclei. The reaction is usually accompanied by the ejection of one or more neutrons.

$$^{252}_{98}Cf \rightarrow ^{140}_{54}Xe + ^{108}_{44}Ru + 4\,^{1}_{0}n$$

For all but the very heaviest isotopes, spontaneous fission is a very slow reaction.

Spontaneous fission of ^{238}U, for example, is almost two million times slower than the rate at which this nuclide undergoes -decay.

NEUTRON-RICH VERSUS NEUTRON-POOR NUCLIDES

In 1934 Enrico Fermi proposed a theory that explained the three forms of beta decay. He argued that a neutron could decay to form a proton by emitting an electron. A proton, on the other hand, could be transformed into a neutron by two pathways. It can capture an electron or it can emit a positron. Electron emission therefore leads to an increase in the atomic number of the nucleus.

$$^{14}_{6}C \rightarrow {}^{14}_{7}N + {}^{0}_{-1}e$$

Both electron capture and positron emission, on the other hand, result in a decrease in the atomic number of the nucleus.

$$^{7}_{4}Be + {}^{0}_{-1}e \rightarrow {}^{7}_{3}Li + h\nu$$

$$^{11}_{6}C \rightarrow {}^{11}_{5}B + {}^{0}_{+1}e$$

A plot of the number of neutrons versus the number of protons for all of the stable naturally occurring isotopes. Several conclusions can be drawn from this plot.

- The stable nuclides lie in a very narrow band of neutron-to-proton ratios.
- The ratio of neutrons to protons in stable nuclides gradually increases as the number of protons in the nucleus increases.
- Light nuclides, such as ^{12}C, contain about the same number of neutrons and protons. Heavy nuclides, such as ^{238}U, contain up to 1.6 times as many neutrons as protons.
- There are no stable nuclides with atomic numbers larger than 83.
- This narrow band of stable nuclei is surrounded by a sea of instability.
- Nuclei that lie above this line have too many neutrons and are therefore neutron-rich.
- Nuclei that lie below this line don't have enough neutrons and are therefore neutron-poor.

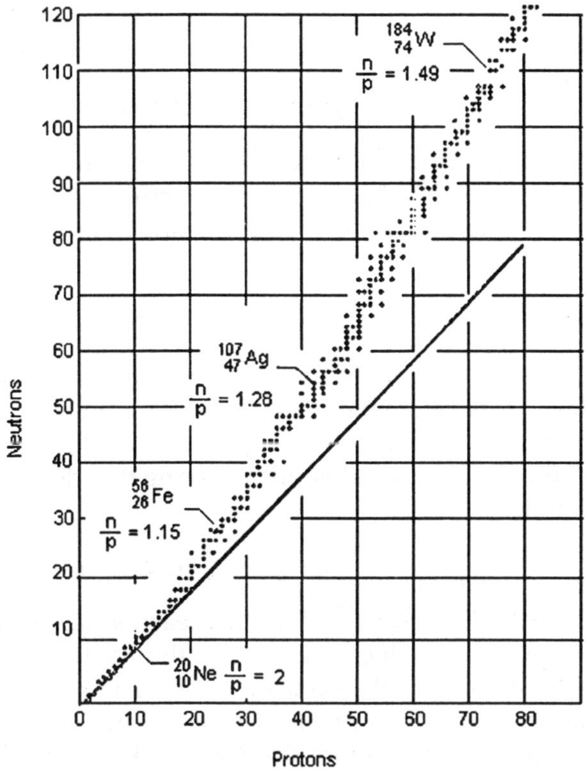

Fig. A Graph of the number of Neutrons Versus the Number of Protons for all Stable Naturally Occurring Nuclei.

The most likely mode of decay for a neutron-rich nucleus is one that converts a neutron into a proton. Every neutron-rich radioactive isotope with an atomic number smaller 83 decays by electron (β^-) emission. ^{14}C, ^{32}P, and ^{35}S, for example, are all neutron-rich nuclei that decay by the emission of an electron.

$$^{35}_{16}S \rightarrow ^{35}_{17}Cl + ^{0}_{-1}e \qquad (\beta^{-}\text{emission})$$

Neutron-poor nuclides decay by modes that convert a proton into a neutron.

Neutron-poor nuclides with atomic numbers less than 83 tend to decay by either electron capture or positron emission. Many of these nuclides decay by both routes, but positron emission is more often observed in the lighter nuclides, such as ^{22}Na.

$$^{22}_{11}Na \rightarrow \, ^{22}_{10}Ne + \, ^{0}_{+1}e \quad (\beta^{+}\text{emission})$$

Electron capture is more common among heavier nuclides, such as ^{125}I, because the 1s electrons are held closer to the nucleus of an atom as the charge on the nucleus increases.

$$^{125}_{53}I + \, ^{0}_{-1} \rightarrow \, ^{125}_{52}Te + h\nu \qquad (\beta^{+}\text{emission})$$

A third mode of decay is observed in neutron-poor nuclides that have atomic numbers larger than 83. Although it is not obvious at first, -decay increases the ratio of neutrons to protons. Consider what happens during the -decay of ^{238}U, for example.

$$^{238}_{92}U \rightarrow \, ^{234}_{90}Th + \, ^{4}_{2}He \qquad (\alpha \text{ decay})$$

The parent nuclide (^{238}U) in this reaction has 92 protons and 146 neutrons, which means that the neutron-to-proton ratio is 1.587. The daughter nuclide (^{234}Th) has 90 protons and 144 neutrons, so its neutron-to-proton ratio is 1.600. The daughter nuclide is therefore slightly less likely to be neutron-pooow.

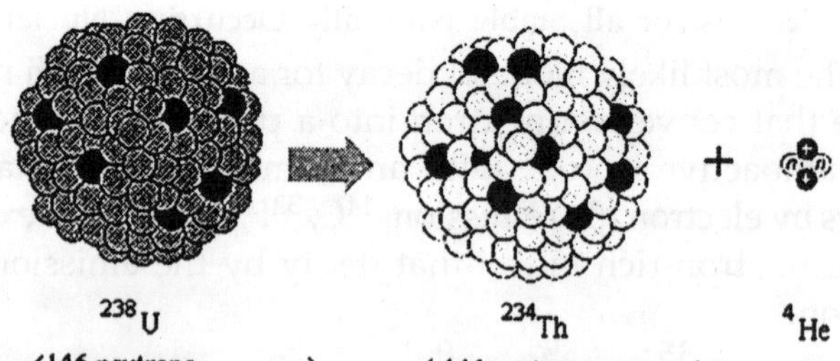

^{238}U $\left(\frac{146 \text{ neutrons}}{92 \text{ protons}} = 1.587\right)$ ^{234}Th $\left(\frac{144 \text{ neutrons}}{90 \text{ protons}} = 1.600\right)$ ^{4}He

BINDING ENERGY CALCULATIONS

We should be able to predict the mass of an atom from the masses of the subatomic particles it contains. A helium atom, for example, contains two protons, two neutrons, and two electrons.

Example: The mass of a helium atom should be 4.0329802 amu.

2(1.0072765) amu = 2.0145530 amu
2(1.0086650) amu = 2.0173300 amu
2(0.0005486) amu = 0.0010972 amu
Total mass = 4.0329802 amu

When the mass of a helium atom is measured, we find that the experimental value is smaller than the predicted mass by 0.0303769 amu.

Predicted mass = 4.0329802 amu
Observed mass = 4.0026033 amu
Mass defect = 0.0303769 amu

The difference between the mass of an atom and the sum of the masses of its protons, neutrons, and electrons is called the mass defect. The mass defect of an atom reflects the stability of the nucleus. It is equal to the energy released when the nucleus is formed from its protons and neutrons. The mass defect is therefore also known as the binding energy of the nucleus.

The binding energy serves the same function for nuclear reactions as H for a chemical reaction. It measures the difference between the stability of the products of the reaction and the starting materials. The larger the binding energy, the more stable the nucleus. The binding energy can also be viewed as the amount of energy it would take to rip the nucleus apart to form isolated neutrons and protons. It is therefore literally the energy that binds together the neutrons and protons in the nucleus.

The binding energy of a nuclide can be calculated from its mass defect with Einstein's equation that relates mass and energy.

$$E = mc^2$$

Example: We found the mass defect of He to be 0.0303769 amu. To obtain the binding energy in units of joules, we must convert the mass defect from atomic mass units to kilograms.

$$0.030796 amu \times \frac{1.6605655 \times 10^{-24}\, g}{1\, amu} \times \frac{1 kg}{1000 g} = 5.04428 \times 10^{-29}\, kg$$

Multiplying the mass defect in kilograms by the square of the speed of light in units of meters per second gives a binding energy for a single helium atom of 4.53358×10^{-12} joules.

$$E = (5.04428 \times 10\text{–}29\ \text{kg})\ (2.99792446 \times 108\ \text{m/s})2$$
$$= 4.53358 \times 10\text{–}12\text{J}$$

Multiplying the result of this calculation by the number of atoms in a mole gives a binding energy for helium of 2.730×10^{12} joules per mole, or 2.730 billion kilojoules per mole.

$$\frac{4.5358 \times 10^{-12}\, J}{1\ \text{atom}} \times \frac{6.022 \times 10^{23}\ \text{atom}}{1\ \text{mol}} = 2.730 \times 10^{12}\ \text{J/mol}$$

This calculation helps us understand the fascination of nuclear reactions. The energy released when natural gas is burned is about 800 kJ/mol. The synthesis of a mole of helium releases 3.4 million times as much energy.

Since most nuclear reactions are carried out on very small samples of material, the mole is not a reasonable basis of measurement. Binding energies are usually expressed in units of electron volts (eV) or million electron volts (MeV) per atom.

Example: The binding energy of helium is 28.3×10^6 eV/ atom or 28.3 MeV/atom.

$$\frac{4.53358 \times 10^{-12}\, J}{1\ \text{atom}} \times \frac{1ev}{1.6021892 \times 10^{-19}\, J} = 28.30 \times 10^6\ \text{ev/atom}$$

Calculations of the binding energy can be simplified by using the following conversion factor between the mass defect in atomic mass units and the binding energy in million electron volts.

$$1\ \text{amu} = 931.5016\ \text{MeV}$$

Binding energies gradually increase with atomic number, although they tend to level off near the end of the periodic table. A more useful quantity is obtained by dividing the binding energy for a nuclide by the total number of protons and neutrons it contains. This quantity is known as the binding energy per nucleon.

The binding energy per nucleon ranges from about 7.5 to 8.8 MeV for most nuclei. It reaches a maximum, however, at an atomic mass of about 60 amu. The largest binding energy per nucleon is observed for ^{56}Fe, which is the most stable nuclide in the periodic table.

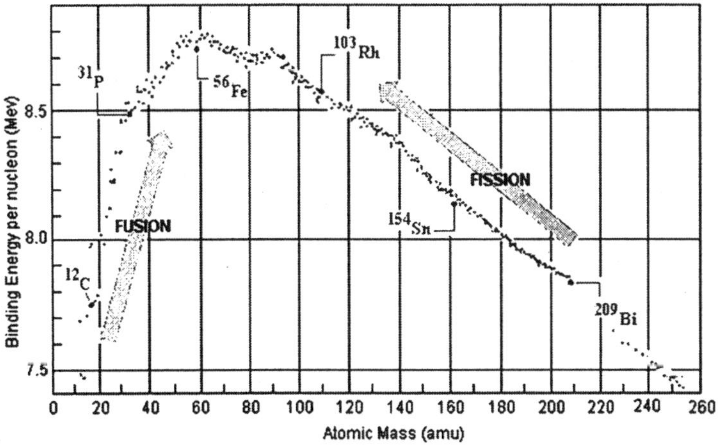

The graph of binding energy per nucleon versus atomic mass explains why energy is released when relatively small nuclei combine to form larger nuclei in fusion reactions.

$$ {}^{12}_{6}C + {}^{12}_{6}C \rightarrow {}^{24}_{12}Mg \text{ (fussion)} $$

It also explains why energy is released when relatively heavy nuclei split apart in fission (literally, "to split or cleave") reactions.

$$ {}^{235}_{92}U \rightarrow {}^{139}_{56}Ba + {}^{94}_{36}Kr + 2\,{}^{1}_{0}n \text{ (fussion)} $$

There are a number of small irregularities in the binding energy curve at the low end of the mass spectrum. The ^{4}He nucleus, for example, is much more stable than its nearest neighbors. The unusual stability of the ^{4}He nucleus explains why -particle decay is usually much faster than the spontaneous fission of a nuclide into two large fragments.

THE KINETICS OF RADIOACTIVE DECAY

Radioactive nuclei decay by first-order kinetics. The rate of radioactive decay is therefore the product of a rate constant (k) times the number of atoms of the isotope in the sample (N).

Rate = kN

The rate of radioactive decay doesn't depend on the chemical state of the isotope. The rate of decay of ^{238}U, for example, is exactly the same in uranium metal and uranium hexafluoride, or any other compound of this element.

The rate at which a radioactive isotope decays is called the activity of the isotope. The most common unit of activity is the curie (Ci), which was originally defined as the number of disintegrations per second in 1 gram of ^{226}Ra.

The curie is now defined as the amount of radioactive isotope necessary to achieve an activity of 3.700×10^{10} disintegrations per second.

The relative rates at which radioactive nuclei decay can be expressed in terms of either the rate constants for the decay or the half-lives of the nuclei. We can conclude that ^{14}C decays more rapidly than ^{238}U, for example, by noting that the rate constant for the decay of ^{14}C is much larger than that for ^{238}U.

^{14}C: $k = 1.210 \times 10^{-4}\ y^{-1}$

^{238}U: $k = 1.54 \times 10^{-10}\ y^{-1}$

We can reach the same conclusion by noting that the half-life for the decay of ^{14}C is much shorter than that for ^{235}U.

^{14}C: $t_{1/2} = 5730\ y$

^{238}U: $t_{1/2} = 4.51 \times 10^{9}\ y$

The half-life for the decay of a radioactive nuclide is the length of time it takes for exactly half of the nuclei in the sample to decay.

In our discussion of the kinetics of chemical reactions, we concluded that the half-life of a first-order process is inversely proportional to the rate constant for this process.

$$t_{1/2} = \frac{\ln 2}{k} = \frac{0.693}{k}$$

The half-life of a nuclide can be used to estimate the amount of a radioactive isotope left after a given number of half-lives.

For more complex calculations, it is easier to convert the half-life of the nuclide into a rate constant and then use the integrated form of the first-order rate law described in the kinetic section.

DATING BY RADIOACTIVE DECAY

The earth is constantly bombarded by cosmic rays emitted by the sun. The total energy received in the form of cosmic rays is small no more than the energy received by the planet from starlight.

But the energy of a single cosmic ray is very large, on the order of several billion electron volts (0?200 million kJ/mol). These highly energetic rays react with atoms in the atmosphere to produce neutrons that then react with nitrogen atoms in the atmosphere to produce ^{14}C.

$$^{14}_{7}N + ^{1}_{0}n \rightarrow ^{14}_{6}C + ^{1}_{2}H$$

The ^{14}C formed in this reaction is a neutron-rich nuclide that decays by electron emission with a half-life of 5730 years.

$$^{14}_{6}C + ^{14}_{7}N + ^{1}_{-1}e$$

Just after World War II, Willard F. Libby proposed a way to use these reactions to estimate the age of carbon-containing substances. The ^{14}C dating technique for which Libby received the Nobel prize was based on the following assumptions.

- ^{14}C is produced in the atmosphere at a more or less constant rate.
- Carbon atoms circulate between the atmosphere, the oceans, and living organisms at a rate very much faster than they decay. As a result, there is a constant concentration of ^{14}C in all living things.
- After death, organisms no longer pick up ^{14}C.

Thus, by comparing the activity of a sample with the activity of living tissue we can estimate how long it has been since the organism *died*.

The natural abundance of ^{14}C is about 1 part in 10^{12} and the average activity of living tissue is 15.3 disintegrations per minute per gram of carbon. Samples used for ^{14}C dating can include charcoal, wood, cloth, paper, sea shells, limestone, flesh, hair, soil, peat, and bone. Since most iron samples also contain carbon, it is possible to estimate the time since iron was last fired by analyzing for ^{14}C.

We now know that one of Libby's assumptions is questionable: The amount of ^{14}C in the atmosphere hasn't been constant with time. Because of changes in solar activity and the earth's magnetic field, it has varied by as much as 5%. More recently, contamination from the burning of fossil fuels and the testing of nuclear weapons has caused significant changes in the amount of radioactive carbon in the atmosphere. Radiocarbon dates are therefore reported in years before the present era (B.P.). By convention, the present era is assumed to begin in 1950, when ^{14}C dating was introduced.

Studies of bristlecone pines allow us to correct for changes in the abundance of ^{14}C with time. These remarkable trees, which grow in the White Mountains of California, can live for up to five thousand years. By studying the ^{14}C activity of samples taken from the annual growth rings in these trees, researchers have developed a calibration curve for ^{14}C dates from the present back to 5145 B.C.

After roughly 45,000 years (eight half-lives), a sample retains only 0.4% of the ^{14}C activity of living tissue. At that point it becomes too old to date by radiocarbon techniques. Other radioactive isotopes can be used to date rocks, soils, or archaeological objects that are much older. Potassium-argon dating, for example, has been used to date samples up to 4.3 billion years old. Naturally occurring potassium contains 0.0118% by weight of the radioactive ^{40}K isotope. This isotope decays to ^{40}Ar with a half-life of 1.3 billion years. The ^{40}Ar produced after a rock crystallizes is trapped in the crystal lattice. It can be released, however, when the rock is melted at temperatures up to 2000C. By measuring the amount of ^{40}Ar released when the rock is melted and comparing it with the amount of potassium in the sample, the time since the rock crystallized can be determined.

IONIZING RADIATION

IONIZING VERSUS NON-IONIZING RADIATION

We live in a sea of radiation. In recent years, people have learned to fear the effects of radiation. They don't want to live near nuclear reactors. They are frightened by reports of links between excess exposure to sunlight and skin cancer. They are afraid of the leakage from microwave ovens, or the radiation produced by their television sets.

Several factors combine to heighten the public's anxiety about both the short-range and long-range effects of radiation. Perhaps the most important source of fear is the fact that radiation can't be detected by the average person. Furthermore, the effects of exposure to radiation might not appear for months or years or even decades.

To understand the biological effects of radiation we must first understand the difference between ionizing radiation and non-ionizing radiation.

In general, two things can happen when radiation is absorbed by matter: excitation or ionization.

- Excitation occurs when the radiation excites the motion of the atoms or molecules, or excites an electron from an occupied orbital into an empty, higher-energy orbital.
- Ionization occurs when the radiation carries enough energy to remove an electron from an atom or molecule.

Because living tissue is 70-90% water by weight, the dividing line between radiation that excites electrons and radiation that forms ions is often assumed to be equal to the ionization of water: 1216 kJ/mol.

Radiation that carries less energy can only excite the water molecule. It is therefore called non-ionizing radiation. Radiation that carries more energy than 1216 kJ/mol can remove an electron from a water molecule, and is therefore called ionizing radiation.

The energies of various kinds of radiation. Radio waves, microwaves, infrared radiation, and visible light are all forms of non-ionizing radiation. X-rays, -rays, and - and β -particles are forms of ionizing radiation.

The dividing line between ionizing and non-ionizing radiation in the electromagnetic spectrum falls in the ultraviolet portion of the spectrum.

It is therefore useful to divide the UV spectrum into two categories: UV_A and UV_B. Radiation at the high-energy end of the UV spectrum can be as dangerous as x-rays or -rays.

Energies of Ionizing and Non-Ionizing Forms of Radiation

Radiation	*Typical Frequency (s^{-1})*	*Typical Energy (kJ/mol)*	
Particles			
α-particles		4.1×10^8	ionizing radiation
β -particles		1.5×10^7	
Electromagnetic radiation			
Cosmic rays	$6 \times 10^{21} s^{-1}$	2.4×10^9	
-rays	$3 \times 10^{20} s^{-1}$	1.2×10^8	
x-rays	$3 \times 10^{17} s^{-1}$	1.2×10^5	
Ultraviolet	$3 \times 10^{15} s^{-1}$	1200	Non-ionizing
Visible	$5 \times 10^{14} s^{-1}$	200	
Infrared	$3 \times 10^{13} s^{-1}$	12	
Microwaves	$3 \times 10^9 s^{-1}$	1.2×10^{-3}	
Radiation			
Radio waves	$3 \times 10^7 s^{-1}$	1.2×10^{-5}	

When ionizing radiation passes through living tissue, electrons are removed from neutral water molecules to produce H_2O^+ ions. Between three and four water molecules are ionized for every 1.6×10^{-17} joules of energy absorbed in the form of ionizing radiation.

$$H_2O \rightarrow H_2O^+ + e^-$$

The H_2O^+ ion should not be confused with the H_3O^+ ion produced when acids dissolve in water. The H_2O^+ ion is an example of a free radical, which contains an unpaired valence-shell electron. Free radicals are extremely reactive.

The radicals formed when ionizing radiation passes through water are among the strongest oxidizing agents that can exist in aqueous solution. At the molecular level, these oxidizing agents destroy biologically active molecules by either removing electrons or removing hydrogen atoms. This often leads to damage to the membrane, nucleus, chromosomes, or mitochondria of the cell that either inhibits cell division, results in cell death, or produces a malignant cell.

BIOLOGICAL EFFECTS OF IONIZING RADIATION

From the time that radioactivity was discovered, it was obvious that it caused damage. As early as 1901, Pierre Curie discovered that a sample of radium placed on his skin produced wounds that were very slow to heal. What some find surprising is the magnitude of the difference between the biological effects of non-ionizing radiation, such as light and microwaves, and ionizing radiation, such as high-energy ultraviolet radiation, x-rays, -rays, and - or β -particles.

Radiation at the low-energy end of the electromagnetic spectrum, such as radio waves and microwaves, excites the movement of atoms and molecules, which is equivalent to heating the sample. Radiation in or near the visible portion of the spectrum excites electrons into higher-energy orbitals. When the electron eventually falls back to a lower-energy state, the excess energy is given off to neighboring molecules in the form of heat. The principal effect of non-ionizing radiation is therefore an increase in the temperature of the system.

We experience the fact that biological systems are sensitive to heat each time we cook with a microwave oven, or spend too long in the sun. But it takes a great deal of non-ionizing radiation to reach dangerous levels. We can assume, for example, that absorption of enough radiation to produce an increase of about 6C in body temperature would be fatal. Since the average 70-kilogram human is 80% water by weight, we can use the heat capacity of water to calculate that it would take about 1.5 million joules of non-ionizing radiation to kill the average human. If this energy was carried by visible light with a frequency of 5 x 10^{14} s^{-1}, it would correspond to absorption of about seven moles of photons.

Ionizing radiation is much more dangerous. A dose of only 300 joules of x-ray or -ray radiation is fatal for the average human, even though this radiation raises the temperature of the body by only 0.001C. -particle radiation is even more dangerous; a dose equivalent to only 15 joules is fatal for the average human. Whereas it takes seven moles of photons of visible light to produce a fatal dose of non-ionizing radiation, absorption of only 7×10^{-10} moles of the -particles emitted by ^{238}U is fatal.

There are three ways of measuring ionizing radiation.

- Measure the activity of the source in units of disintegrations per second or curies, which is the easiest measurement to make.
- Measure the radiation to which an object is exposed in units of roentgens by measuring the amount of ionization produced when this radiation passes through a sample of air.
- Measure the radiation absorbed by the object in units of radiation absorbed doses or "rads."

  This is the most useful quantity, but it is the hardest to obtain.

One radiation absorbed dose, or rad, corresponds to the absorption of 10^{-5} joules of energy per gram of body weight. Because this is equivalent to 0.01 J/kg, one rad produces an increase in body temperature of about 2×10^{-6}C.

At first glance, the rad may seem to be a negligibly small unit of measurement.

The destructive power of the radicals produced when water is ionized is so large, however, that cells are inactivated at a dose of 100 rads, and a dose of 400 to 450 rads is fatal for the average human.

Not all forms of radiation have the same efficiency for damaging biological organisms. The faster energy is lost as the radiation passes through the tissue, the more damage it does.

To correct for the differences in radiation biological effectiveness (RBE) among various forms of radiation, a second unit of absorbed dose has been defined. The roentgen equivalent man, or rem, is the absorbed dose in rads times the biological effectiveness of the radiation.

$$\text{rems} = \text{rads} \times \text{RBE}$$

Values for the RBE of different forms of radiation are given.

The Radiation Biological Effectiveness of Various Forms of Radiation

Radiation	*RBE*
x-rays and -rays	1
β - particles with energies larger than 0.03 MeV	1
β - particles with energies less than 0.3 MeV	1.7
thermal (slow-moving) neutrons	3
fast-moving neutrons or protons	10
α-particles or heavy ions	20

Estimates of the per capita exposure to radiation . These estimates include both external and internal sources of natural background radiation.

Average Whole-Body Exposure Levels for Sources of Ionizing Radiation

Source	*Per Capita Dose (rems / y)*
natural background	0.082
medical x-rays	0.077
nuclear test fallout	0.005
consumer and industrial products	0.005
nuclear power industry	0.001
	total: 0.170

External sources include cosmic rays from the sun and -particles or -rays emitted from rocks and soil. Internal sources include nuclides that enter the body when we breathe (^{14}C, ^{85}Kr, ^{220}Rn, and ^{222}Rn) and through the food chain (^{3}H, ^{14}C, ^{40}K, ^{90}Sr, ^{131}I, and ^{137}Cs). The actual dose from natural radiation depends on where one lives. People who live in the Rocky Mountains, for example, receive twice as much background radiation as the national average because there is less atmosphere to filter out the cosmic rays from the sun. The average dose from medical x-rays has decreased in recent years because of advances in the sensitivity of the photographic film used for x-rays. Radiation from nuclear test fallout has also decreased as a result of the atmospheric nuclear test ban. The threat of fallout from the testing of nuclear weapons can be appreciated by noting that a Chinese atmospheric test in 1976 led to the contamination of milk in the Harrisburg, PA, vicinity at a level of 300 pCi (3.00×10^{-10} Ci) per lite.

This was about eight times the level of contamination (41 pCi per liter) that resulted from the accident at Three Mile Island.

The contribution to the radiation absorbed dose from consumer and industrial products includes radiation from construction materials, x-rays emitted by television sets, and inhaled tobacco smoke.

The most recent estimate of the total radiation emitted from the mining and milling of uranium, the fabrication of reactor fuels, the storage of radioactive wastes, and the operation of nuclear reactors is less than 0.001 rem per year. The total dose from ionizing radiation for the average American is about 0.170 rem per year.

The Committee on the Biological Effects of Ionizing Radiation of the National Academy of Sciences recently estimated that an increase in this dose to a level of 1 rem per year would result in 169 additional deaths from cancer per million people exposed.

This can be compared with the 170,000 cancer deaths that would normally occur in a population this size that was not exposed to this level of radiation.

The principal effect of low doses of ionizing radiation is to induce cancers, which may take up to 20 years to develop. What is the effect of high doses of ionizing radiation? Cells that are actively dividing are more sensitive to radiation than cells that aren't.

Thus, cells in the liver, kidney, muscle, brain, and bone are more resistant to radiation than the cells of bone marrow, the reproductive organs, the epithelium of the intestine, and the skin, which suffer the most damage from radiation. Damage to the bone marrow is the main cause of death at moderately high levels of exposure (200 to 1000 rads).

Damage to the gastrointestinal tract is the major cause of death for exposures on the order of 100 to 10,000 rads. Massive damage to the central nervous system is the cause of death from extremely high exposures (over 10,000 rads).

NATURAL VERSUS INDUCED RADIOACTIVITY

NATURAL RADIOACTIVITY

The vast majority of the nuclides found in nature are stable. If our planet is 4.6 billion years old, the only radioactive isotopes that should remain are members of three classes.

- Isotopes with half-lives of at least 10^9 years, such as ^{238}U.
- Daughter nuclides produced when long-lived radioactive nuclides decay, such as the ^{234}Th ($t_{1/2}$ = 24.1 days) produced by the -decay of ^{238}U.
- Nuclides such as ^{14}C that are still being synthesized.

One factor that influences the stability of a nuclide is the ratio of neutrons to protons. (Nuclei that contain either too many or too few neutrons are unstable.) Another factor that affects the stability of nuclides can be understood by examining patterns in the numbers of protons and neutrons in stable nuclides.

Half of the elements in the periodic table must have an odd number of protons because atomic numbers that are odd are just as likely to occur as those that are even. In spite of this, about 80% of the stable nuclides have an even number of protons. Very few elements with an odd atomic number have more than one stable isotope. Stable isotopes abound, however, among elements with even atomic numbers. Ten stable isotopes are known for tin (Z = 50), for example. It is also interesting to note that 91% of the stable isotopes of elements with an odd number of protons have an even number of neutrons. These observations suggest that certain combinations of protons and neutrons are particularly stable.

There are magic numbers of electrons. Electron configurations with 2, 10, 18, 36, 54, and 86 electrons are unusually stable. There also seem to be magic numbers of neutrons and protons. Nuclei with 2, 8, 20, 28, 50, 82, or 126 protons or neutrons are unusually stable. This observation explains the anomalously large binding energies observed for ^4He, ^{16}O, and ^{20}Ne. In each case, the nuclide has an even number of both protons and neutrons. ^{20}Ne has a magic number of nucleons when both protons and neutrons are counted. ^{4}He and ^{16}O have magic numbers of both protons and neutrons. The resulting stability of the ^{4}He nucleus might explain why so many heavy nuclei undergo -particle decay by ejecting an $^{4}He^{2+}$ ion from the nucleus of the atom.

If nuclei tend to be more stable when they have even numbers of protons and neutrons, it isn't surprising that nuclides with an odd number of both protons and neutrons are unstable. ^{40}K is one of only five naturally occurring nuclides that contain both an odd number of protons and an odd number of neutrons. This nuclide simultaneously undergoes the electron capture and positron emission expected for neutron-poor nuclides and the electron emission observed with neutron-rich nuclides.

Only 18 radioactive isotopes with atomic numbers of 80 or less can be found in nature. With the exception of ^{14}C, which is continuously synthesized in the atmosphere, all these elements have lifetimes longer than 10^9 years. Although these isotopes all undergo radioactive decay, they decay so slowly that reasonable quantities are still present, 4.6 billion years after the planet was formed.

Another 45 natural radioactive isotopes have atomic numbers larger than 80. These nuclides fall into three families. The parent nuclide is ^{232}Th, which undergoes -decay to form ^{228}Ra. The product of this reaction decays by β -emission to form ^{228}Ac, which decays to ^{228}Th, and so on, until the stable ^{208}Pb isotope is formed. This family of radionuclides is called the 4n series, because all its members have a mass number that can be divided by 4.

A second family of radioactive nuclei starts with ^{238}U and decays to form the stable ^{206}Pb isotope. Every member of this series has a mass number that fits the equation 4n + 2. The third family, known as the 4n + 3 series, starts with ^{235}U and decays to ^{207}Pb. A 4n + 1 series once existed, which started with ^{237}Np and decayed to form the only stable isotope of bismuth, ^{209}Bi.

The half-life of every member of this series is less than 2 x 10^6 years, however, so none of the nuclides produced by the decay of neptunium remain in detectable quantities on the earth.

INDUCED RADIOACTIVITY

In 1934, Irene Curie, the daughter of Pierre and Marie Curie, and her husband, Frederic Joliot, announced the first synthesis of an artificial radioactive isotope. They bombarded a thin piece of aluminum foil with -particles produced by the decay of polonium and found that the aluminum target became radioactive. Chemical analysis showed that the product of this reaction was an isotope of phosphorus.

$$ {}^{27}_{12}Al + {}^{4}_{2}He \rightarrow {}^{30}_{15}P + {}^{1}_{0}n $$

In the next 50 years, more than 2000 other artificial radionuclides were synthesized.

A shorthand notation has been developed for nuclear reactions such as the reaction discovered by Curie and Joliot. The parent (or target) nuclide and the daughter nuclide are separated by parentheses that contain the symbols for the particle that hits the target and the particle or particles released in this reaction.

$$ {}^{27}_{13}Al(\alpha n){}^{30}_{15}P $$

The nuclear reactions used to synthesize artificial radionuclides are characterized by enormous activation energies. Three devices are used to overcome these activation energies: linear accelerators, cyclotrons, and nuclear reactors. Linear accelerators or cyclotrons can be used to excite charged particles such as protons, electrons, -particles, or even heavier ions, which are then focused on a stationary target. The following reaction, for example, can be induced by a cyclotron or linear accelerator.

$$ {}^{24}_{12}Mg + {}^{2}_{1}H \rightarrow {}^{22}_{11}Na + {}^{4}_{2}He $$

Because these reactions involve the capture of a positively charged particles, they usually produce a neutron-poor nuclide.

Artificial radionuclides are also synthesized in nuclear reactors, which are excellent sources of slow-moving, or thermal neutrons. The absorption of a neutron usually results in a neutron-rich nuclide. The following neutron absorption reaction occurs in the cooling systems of nuclear reactors cooled with liquid sodium metal.

$$ {}^{23}_{11}Na + {}^{1}_{0}n \rightarrow {}^{24}_{11}Na + \gamma $$

In 1940, absorption of thermal neutrons was used to synthesize the first elements with atomic numbers larger than the heaviest naturally occurring element, uranium. The first of these truly artificial elements were neptenium and plutonium, which were synthesized by McMillan and Abelson by irradiating ^{238}U with neutrons to form ^{239}U

$$ {}^{238}_{92}U + {}^{1}_{0}n \rightarrow {}^{239}_{92}U + \gamma \quad \text{(Neutron capture)} $$

which undergoes β^--decay to form ^{239}Np and then ^{239}Pu.

$$ {}^{238}_{92} \rightarrow {}^{239}_{92}Np + {}^{0}_{-1}e $$

$$ {}^{238}_{93} \rightarrow {}^{239}_{94}Pu + {}^{0}_{-1}e $$

Larger bombarding particles were eventually used to produce even heavier transuranium elements.

$$ {}^{253}_{99}Es + {}^{4}_{2}He \rightarrow {}^{256}_{101}Md + {}^{1}_{0}n $$

$$ {}^{246}_{96}Cm + {}^{12}_{6}C \rightarrow {}^{254}_{102}No + 4\,{}^{1}_{0}n $$

The half-lives for -decay and spontaneous fission decrease as the atomic number of the element increases. Element 104, for example, has a half-life for spontaneous fission of 0.3 seconds. Elements therefore become harder to characterize as the atomic number increases. Recent theoretical work has predicted that a magic number of protons might exist at Z = 114. This work suggests that there is an island of stability in the sea of unstable nuclides.

If this theory is correct, superheavy elements could be formed if we could find a way to cross the gap between elements Z = 109 through Z = 114.

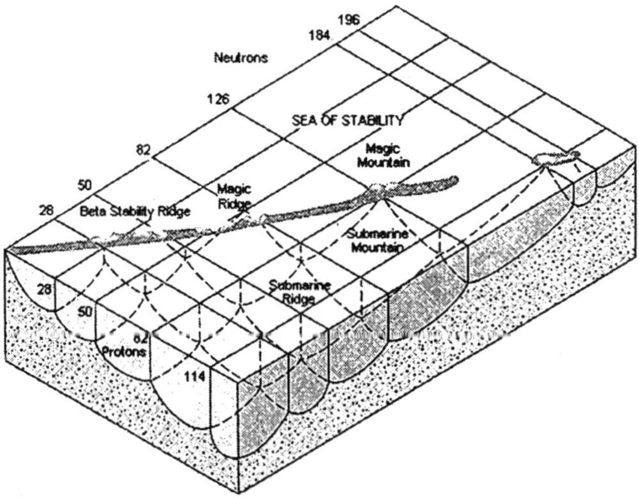

There is some debate about the number of neutrons needed to overcome the proton-proton repulsion in a nucleus with 114 protons. The best estimates suggest that at least 184 neutrons, and perhaps as many as 196, would be needed. It is not an easy task to bring together two particles that give both the correct number of total protons and the necessary neutrons to produce a nuclide with a half-life long enough to be detected. If we start with a relatively long-lived parent nuclide, such as ^{251}Cf ($t_{1/2}$ = 800 y) and bombard this nucleus with a heavy ion, such as ^{32}S, we can envision producing a daughter nuclide with the correct atomic number, but the mass number would be too small by at least 16 amu.

$$^{239}_{92}U \rightarrow {}^{239}_{93}Np + {}^{0}_{-1}e$$

$$^{239}_{93}Np \rightarrow {}^{239}_{94}Pu + {}^{0}_{-1}e \quad {}^{251}_{98}Cf + {}^{32}_{16}S \rightarrow {}^{282}_{114}X + {}^{1}_{0}n$$

An expanded periodic table for elements up to Z = 168 . Elements 104 through 112 are transition metals that fill the 6d orbitals. Elements 113 through 120 are main-group elements in which the 7p and 8s orbitals are filled.

The next subshell is the 5g atomic orbital, which can hold up to 18 electrons. There is reason to believe that the 5g and 6f orbitals will be filled at the same time. The next 32 elements are therefore grouped into a so-called superactinide series.

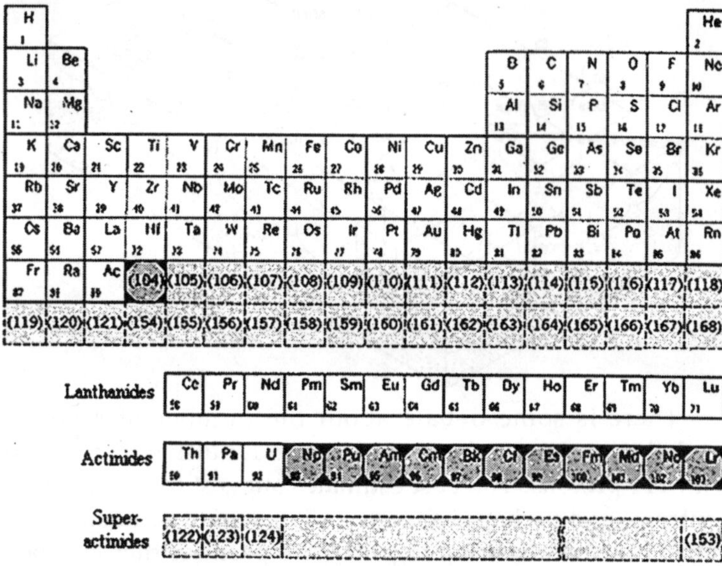

Fig. An Extended Version of the Periodic Table that Predicts the Positions for all Elements up to Atomic Number 168.

NUCLEAR FISSION AND NUCLEAR FUSION

NUCLEAR FISSION

The graph of binding energy per nucleon suggests that nuclides with a mass larger than about 130 amu should spontaneously split apart to form lighter, more stable, nuclides. Experimentally, we find that spontaneous fission reactions occur for only the very heaviest nuclides those with mass numbers of 230 or more.

Even when they do occur, these reactions are often very slow. The half-life for the spontaneous fission of ^{238}U, for example, is 10^{16} years, or about two million times longer than the age of our planet!

We don't have to wait, however, for slow spontaneous fission reactions to occur. By irradiating samples of heavy nuclides with slow-moving thermal neutrons it is possible to induce fission reactions. When ^{235}U absorbs a thermal neutron, for example, it splits into two particles of uneven mass and releases an average of 2.5 neutrons.

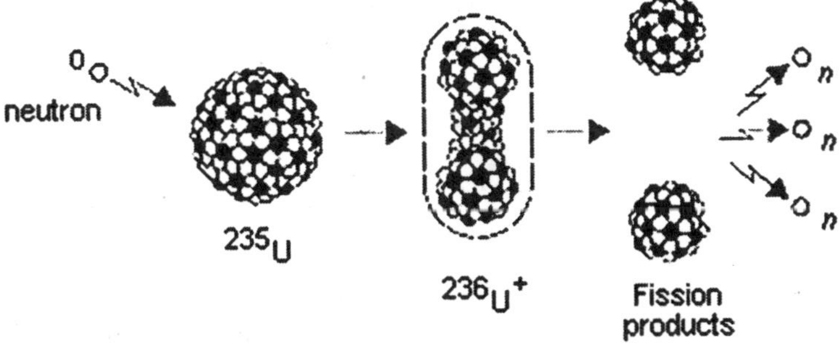

Fig. The Absorption of a Neutron by ^{238}U induces Oscillations in the Nucleus that deform it until it splits into Fragments the way a Drop of Liquid might break into smaller Droplets.

More than 370 daughter nuclides with atomic masses between 72 and 161 amu are formed in the thermal-neutron-induced fission of ^{235}U, including the two products shown below.

$$^{235}_{92}U + ^{1}_{0}n \rightarrow ^{139}_{56}Ba + ^{94}_{36}Kr + 3\,^{1}_{0}n$$

Several isotopes of uranium undergo induced fission. But the only naturally occurring isotope in which we can induce fission with thermal neutrons is ^{235}U, which is present at an abundance of only 0.72%. The induced fission of this isotope releases an average of 200 MeV per atom, or 80 million kilojoules per gram of ^{235}U. The attraction of nuclear fission as a source of power can be understood by comparing this value with the 50 kJ/g released when natural gas is burned.

The first artificial nuclear reactor was built by Enrico Fermi and co-workers beneath the University of Chicago's football stadium and brought on line on December 2, 1942.

This reactor, which produced several kilowatts of power, consisted of a pile of graphite blocks weighing 385 tons stacked in layers around a cubical array of 40 tons of uranium metal and uranium oxide. Spontaneous fission of ^{238}U or ^{235}U in this reactor produced a very small number of neutrons. But enough uranium was present so that one of these neutrons induced the fission of a ^{235}U nucleus, thereby releasing an average of 2.5 neutrons, which catalyzed the fission of additional ^{235}U nuclei in a chain reaction. The amount of fissionable material necessary for the chain reaction to sustain itself is called the critical mass.

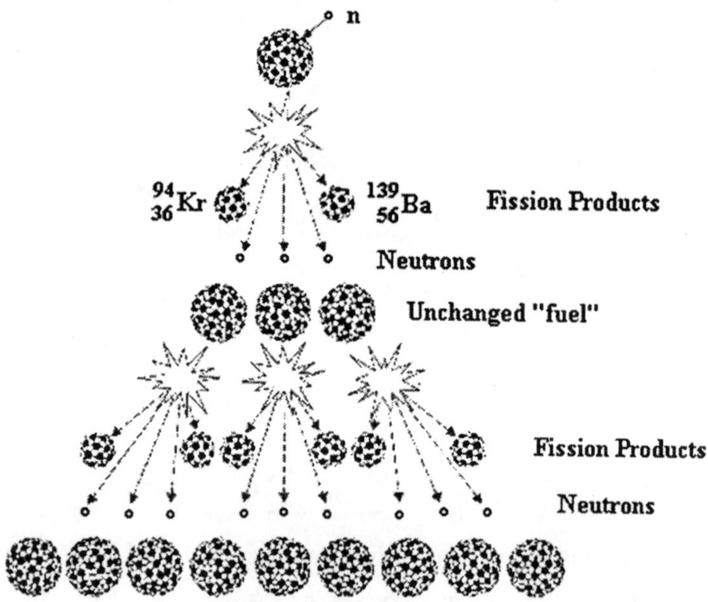

Fig. Fission Reaction.

The Fermi reactor at Chicago served as a prototype for larger reactors constructed in 1943 at Oak Ridge, Tennessee, and Hanford, Washington, to produce ^{239}Pu for one of the atomic bombs dropped on Japan at the end of World War II. As we have seen, some of the neutrons released in the chain reaction are absorbed by ^{238}U to form ^{239}U, which undergoes decay by the successive loss of two β -particles to form ^{239}Pu. ^{238}U is an example of a fertile nuclide. It doesn't undergo fission with thermal neutrons, but it can be converted to ^{239}Pu, which does undergo thermal-neutron-induced fission.

Fission reactors can be designed to handle naturally abundant ^{235}U, as well as fuels described as slightly enriched (2-5% ^{235}U), highly enriched (20-30% ^{235}U), or fully enriched (>90% ^{235}U). Heat generated in the reactor core is transferred to a cooling agent in a closed system. The cooling agent is then passed through a series of heat exchangers in which water is heated to steam. The steam produced in these exchangers then drives a turbine that generates electrical power. There are two ways of specifying the power of such a plant: the thermal energy produced by the reactor or the electrical energy generated by the turbines. The electrical capacity of the plant is usually about one third of the thermal power.

It takes 10^{11} fissions per second to produce one watt of electrical power. As a result, about one gram of fuel is consumed per day per megawatt of electrical energy produced. This means that one gram of waste products is produced per megawatt per day, which includes 0.5 grams of ^{239}Pu. These waste products must be either reprocessed to generate more fuel or stored for the tens of thousands of years it takes for the level of radiation to reach a safe limit.

If we could design a reactor in which the ratio of the ^{239}Pu or ^{233}U produced to the ^{235}U consumed was greater than 1, the reactor would generate more fuel than it consumed. Such reactors are known as breeders, and commercial breeder reactors are now operating in France.

The key to an efficient breeder reactor is a fuel that gives the largest possible number of neutrons released per neutron absorbed. The breeder reactors being built today use a mixture of PuO_2 and UO_2 as the fuel and fast neutrons to activate fission.

Fast neutrons carry an energy of at least several KeV and therefore travel 10,000 or more times faster than thermal neutrons. ^{239}Pu in the fuel assembly absorbs one of these fast neutrons and undergoes fission with the release of three neutrons. ^{238}U in the fuel then captures one of these neutrons to produce additional ^{239}Pu.

The advantage of breeder reactors is obviousthey mean a limitless supply of fuel for nuclear reactors. There are significant disadvantages, however. Breeder reactors are more expensive to build. They are also useless without a subsidiary industry to collect the fuel, process it, and ship the ^{239}Pu to new reactors.

It is the reprocessing of ^{239}Pu that concerns most of the critics of breeder reactors. ^{239}Pu is so dangerous as a carcinogen that the nuclear industry places a limit on exposure to this material that assumes workers inhale no more than 0.2 micrograms of plutonium over their lifetimes. There is also concern that the ^{239}Pu produced by these reactors might be stolen and assembled into bombs by terrorist organizations.

The fate of breeder reactors in the United States is linked to economic considerations. Because of the costs of building these reactors and safely reprocessing the ^{239}Pu produced, the breeder reactor becomes economical only when the scarcity of uranium drives its price so high that the breeder reactor becomes cost effective by comparison. If nuclear energy is to play a dominant role in the generation of electrical energy in the 21st century, breeder reactors eventually may be essential.

Although the "pile" Fermi constructed at the University of Chicago in 1942 was the first artificial nuclear reactor, it was not the first fission reactor to exist on Earth. In 1972, a group of French scientists discovered that uranium ore from a deposit in the Oklo mine in Gabon, West Africa, contained 0.4% ^{235}U instead of the 0.72% abundance found in all other sources of this ore.

Analysis of the trace elements in the ore suggested that the amount of ^{235}U in this ore was unusually small because natural fission reactors operated in this deposit for a period of 600,000 to 800,000 years about 2 billion years ago.

NUCLEAR FUSION

The graph of binding energy per nucleon suggests another way of obtaining useful energy from nuclear reactions. Fusing two light nuclei can liberate as much energy as the fission of ^{235}U or ^{239}Pu. The fusion of four protons to form a helium nucleus, two positrons (and two neutrinos), for example, generates 24.7 MeV of energy.

$$4\,{}^{1}_{1}H \rightarrow {}^{4}_{2}He + 2\,{}^{0}_{+1}e$$

Most of the energy radiated from the surface of the sun is produced by the fusion of protons to form helium atoms within its core.

Fusion reactions have been duplicated in man-made devices. The enormous destructive power of the ^{235}U-fueled atomic bomb dropped on Hiroshima on August 6, 1945, which killed 75,000 people, and the ^{239}Pu-fueled bomb dropped on Nagasaki three days later touched off a violent debate after World War II about the building of the next superweapon a fusion, or "hydrogen," bomb. Alumni of the Manhattan project, who had developed the atomic bomb, were divided on the issue.

Ernest Lawrence and Edward Teller fought for the construction of the fusion device. J. Robert Oppenheimer and Enrico Fermi argued against it. The decision was made to develop the weapon, and the first artificial fusion reaction occurred when the hydrogen bomb was tested in November 1952.

The history of fusion research is therefore the opposite of fission research. With fission, the reactor came first, and then the bomb was built. With fusion, the bomb was built long before any progress was made toward the construction of a controlled fusion reactor. More than 40 years after the first hydrogen bomb was exploded, the feasibility of controlled fusion reactions is still open to debate. The reaction that is most likely to fuel the first fusion reactor is the thermonuclear d-t, or deuterium-tritium, reaction. This reaction fuses two isotopes of hydrogen, deuterium (^{2}H) and tritium (^{3}H), to form helium and a neutron.

$${}^{2}_{1}H + {}^{3}_{1}H \rightarrow {}^{2}_{2}He + {}^{1}_{0}n$$

If we consider the implications of this reaction we can begin to understand why it is called a thermonuclear reaction and why it is so difficult to produce in a controlled manner. The d-t reaction requires that we fuse two positively charged particles. This means that we must provide enough energy to overcome the force of repulsion between these particles before fusion can occur. To produce a self-sustaining reaction, we have to provide the particles with enough thermal energy so that they can fuse when they collide.

Each fusion reaction is characterized by a specific ignition temperature, which must be surpassed before the reaction can occur. The d-t reaction has an ignition temperature above 10^8 K. In a hydrogen bomb, a fission reaction produced by a small atomic bomb is used to heat the contents to the temperature required to initiate fusion. Obtaining the same result in a controlled reaction is much more challenging.

Any substance at temperatures approaching 10^8 K will exist as a completely ionized gas, or plasma. The goals of fusion research at present include the following.

- To achieve the required temperature to ignite the fusion reaction.
- To keep the plasma together at this temperature long enough to get useful amounts of energy out of the thermonuclear fusion reactions.
- To obtain more energy from the thermonuclear reactions than is used to heat the plasma to the ignition temperature.

These are not trivial goals. The only reasonable container for a plasma at 10^8 K is a magnetic field. Both doughnut-shaped (toroidal) and linear magnetic bottles have been proposed as fusion reactors. But reactors that produce high enough temperatures for ignition are not the same as the reactors that have produced long enough confinement times for the plasma to provide useful amounts of energy.

A second approach to a controlled fusion reactor involves hitting fuel pellets containing the proper reagents for the thermonuclear reaction with pulsed beams of laser power. If enough power was delivered, the fuel pellets would collapse upon themselves, or implode, to reach densities several orders of magnitude greater than normal. This could produce a plasma both hot enough and dense enough to initiate fusion reactions.

NUCLEAR SYNTHESIS AND NUCLEAR MEDICINE

NUCLEAR SYNTHESIS

Within the last 60 years, scientists have generated a new story of creation that offers a model for understanding how nuclei are synthesized.

In 1929, Edwin Hubble provided evidence to suggest that our universe is expanding. Between 1946 and 1948, George Gamow and co-workers generated a model which assumed that the primordial substance, or ylem, from which all other matter was created was an extraordinarily hot, dense singularity that exploded in a "Big Bang" and has been expanding ever since. This model assumes that neutrons in the ylem were transformed into protons by β -decay. Neutrons and protons then combined to form ^4He atoms before the temperature and pressure of the fireball decayed to the point at which no further nuclear reactions were possible.

The first-generation stars condensed out of this cloud of hydrogen and helium. As the gas condensed by gravitational attraction, it became warmer. Eventually, the temperature reached 10^7 K, and first-generation, main-sequence stars were born. The temperature at the cores of these stars was high enough to ignite the following thermonuclear reactions.

$$^{1}_{1}H + ^{1}_{1}H \rightarrow ^{2}_{1}H + ^{1}_{0}e$$

$$^{1}_{1}H + ^{2}_{1}H \rightarrow ^{3}_{2}He$$

$$^{3}_{2}He + ^{3}_{2}He \rightarrow ^{4}_{2}He + 2\,^{1}_{1}H$$

The net result of these reactions is the formation of a helium atom from four protons.

$$4\,^{1}_{1}H \rightarrow ^{4}_{2}He + ^{0}_{+1}e$$

Eventually the heat generated in this reaction was enough to halt the gravitational collapse of the star, which entered a stable period during which the energy generated by this reaction balanced the energy radiated at the surface.

The hydrogen-burning reactions in a main-sequence star are concentrated in the core. When enough hydrogen has been consumed, the core begins to collapse, and the temperature of the core rises above 10^8 K. (The larger the star, the more rapidly it radiates energy from its surface, and the more rapidly it consumes the hydrogen in the core).

As the core collapses, the hydrogen-containing outer shell expands, and the surface of the star cools. Stars that have reached this point in their evolution include the so-called red giants.

Temperatures in the core of these red giants are high enough to ignite further thermonuclear fusion reactions, such as the following.

$$3\,{}^{4}_{2}He \rightarrow {}^{12}_{6}C$$

This reaction becomes the principal source of energy in a red giant, although there is undoubtedly some burning of hydrogen to helium in the outer shell of the star. As the amount of ^{12}C in the core increases, further reactions occur to form ^{16}O and ^{20}Ne.

$${}^{12}_{6}C + {}^{4}_{2} \rightarrow {}^{16}_{8}O + \gamma$$

$${}^{16}_{8}O + {}^{4}_{2}He \rightarrow {}^{20}_{10}Ne + \gamma$$

Eventually, the helium in the core is exhausted, and the core collapses further, reaching temperatures of 6-7 x 10^8 K. At this point, more complex reactions take place that produce nuclides such as ^{28}Si and ^{32}S.

$${}^{12}_{6}C + {}^{16}_{8}O \rightarrow {}^{28}_{14}Si + \gamma$$

$${}^{16}_{8}O + {}^{16}_{8}O \rightarrow {}^{32}_{16}S + \gamma$$

Further gravitational collapse heats the core to temperatures above 10^9 K, and a complex sequence of reactions takes place to synthesize the nuclei with the highest binding energies, such as Fe and Ni.

If the star explodes in a supernova, its contents are ejected across space. Second-generation stars that condense in this region contain not only hydrogen and helium but elements with higher atomic number.

The best estimates of the age of the Milky Way suggest that our galaxy is about 15 billion years old. Our sun and its planets, however, are only 4.6 billion years old. This suggests that the sun is a second-generation star. In such stars, the transformation of hydrogen to helium can be catalyzed by ^{12}C.

$$ {}^{12}_{6}C + {}^{1}_{1}H \rightarrow {}^{13}_{7}N + \gamma $$

$$ {}^{13}_{7}N \rightarrow {}^{13}_{6}C + {}^{0}_{+1} $$

$$ {}^{13}_{6}C + {}^{1}_{1}H \rightarrow {}^{14}_{7}N + \gamma $$

$$ {}^{14}_{6}N + {}^{1}_{1}H \rightarrow {}^{15}_{8}N + \gamma $$

$$ {}^{18}_{8}O \rightarrow {}^{15}_{7}N + {}^{0}_{+1}e $$

$$ {}^{15}_{7}N + {}^{1}_{1}H \rightarrow {}^{12}_{6}C + {}^{4}_{2}He $$

Second-generation stars use a different mechanism to synthesis helium than first-generation stars. They use a sequence of reactions, such as the those shown here, that are catalyzed by isotopes heavier than helium.

Two processes can synthesize elements with atomic numbers larger than that of iron. One of them is relatively slow (s-process), the other is very rapid (r-process). Since the only way to synthesize nuclei with atomic numbers larger than iron is by the absorption of neutrons, both the s-process and the r-process result from (n,) reactions.

In the s-process, neutrons are captured one at a time to form a neutron-rich nuclide that has enough time to undergo - or β -decay before another neutron can be absorbed. An example of an s-process sequence of reactions starts with ^{120}Sn. The capture of a neutron produces ^{121}Sn, which undergoes β^--decay.

If β^--decay occurs before this nuclide captures another neutron, a stable isotope of antimony is formed. Eventually, ^{121}Sb captures a neutron to produce ^{122}Sb, which is transformed into ^{122}Te by β -decay. ^{122}Te can undergo β -decay to form ^{122}I, or it can capture a neutron to form ^{123}Te. With ^{123}Te, we encounter a series of stable isotopes of tellurium. Thus, neutrons are slowly absorbed, one at a time, until we reach ^{127}Te, which decays to ^{127}I, the most abundant isotope of iodine.

$$\begin{matrix}120\\60\end{matrix} Sn(n,\gamma) \begin{matrix}121\\50\end{matrix} Sn \overset{\beta^-}{\rightarrow} \begin{matrix}121\\51\end{matrix} Sb(n,\gamma)$$

$$\begin{matrix}122\\51\end{matrix} Sb \overset{\beta^-}{\rightarrow} \begin{matrix}122\\52\end{matrix} Te(5n,5y) \begin{matrix}127\\52\end{matrix} Te \overset{\beta^-}{\rightarrow} \begin{matrix}127\\53\end{matrix} I$$

This slow process can't account for very heavy nuclides, such as ^{232}Th and ^{238}U, because the lifetimes of the intermediate nuclei with atomic numbers between 83 and 90 are too short for this step-by-step absorption of neutrons to proceed.

Synthesizing appreciable quantities of uranium and thorium requires a rapid process. In the r-process, a number of neutrons are captured in rapid succession, before there is time for - or β -decay to take place. Achieving an r-process reaction, however, requires a very high neutron flux. (These reactions occur during nuclear explosions, for example.) The neutron flux needed to fuel such reactions is not likely to occur in a normal star.

During the moment when a star explodes as a supernova, however, the conditions are ripe for r-process reactions. The heavier elements on this planet were therefore produced in a series of supernova explosions that occurred in this portion of the galaxy before our solar system condensed.

NUCLEAR MEDICINE

Ever since the first x-ray images were obtained by Roentgen in 1895, ionizing radiation and radio nuclides have played a vital role in medicine. This work has been so fruitful that a separate field known as nuclear medicine has developed. Research in this field focuses on either therapeutic or diagnostic uses of radiation.

There are three standard approaches to fighting cancer: surgery, chemotherapy, and radiation. Surgery, by its very nature, is invasive. Chemotherapy and classic approaches to radiation therapy are not selective. Research in recent years has therefore examined new approaches to radiation therapy that specifically attack tumor cells, without damaging normal tissue. The technique known as boron neutron capture therapy provides an example of this work.

Naturally occurring boron consists of two stable isotopes: ^{10}B (19.7%) and ^{11}B (80.3%). ^{10}B absorbs thermal neutrons to form ^{11}B in a nuclear excited state. Although ^{11}B in its nuclear ground state is stable, this excited ^{11}B nuclide undergoes fission to produce ^{7}Li and an α-particle.

$$^{10}_{5}B + ^{1}_{0}n \rightarrow ^{7}_{3}Li + ^{4}_{2}He$$

Because the energy of a thermal neutron is only about 0.025 eV, the neutrons that are not absorbed do relatively little damage to the normal tissue. The -particle emitted in this reaction has an energy of 2.79 MeV, however, which makes it an extremely lethal form of radiation. The RBE for -particle radiation is larger than most other particles because this relatively massive particle loses energy very efficiently as it collides with matter. Radiation damage from the -particle is therefore restricted to the immediate vicinity of the tissue that absorbed the thermal neutron.

Although other common nuclides in living tissue can absorb thermal neutrons, the ability of ^{10}B to absorb thermal neutrons is three orders of magnitude larger than these nuclides. The units with which this measurement is made can be understood by thinking about the area around the nuclide through which the neutron can pass and still be absorbed. The neutron-capture cross section for a hydrogen atom corresponds to a circle with a radius of about 2 x 10^{-13} cm. For a nitrogen atom, the radius of this circle is about 10^{-12} cm. Boron has a neutron-capture cross section that would be described by a circle with a radius of about 2 x 10^{-9} cm. Virtually all of the neutron capture that occurs is therefore concentrated in the tissue that contains boron.

The potential of boron neutron capture therapy was recognized as early as 1936. Early clinical trials in the late 1950s and early 1960s failed to prolong the lives of patients suffering from brain tumors because the boron compounds available for testing did not concentrate selectively in the tumor cells. Recent research has discovered boron-labeled compounds that are sufficiently selective for BNCT to become useful as an adjunct to surgery, or in place of surgery with inoperable cases.

Chapter 5

Quantum Mechanics

Quantum mechanics is a physical science dealing with the behaviour of matter and energy on the scale of atoms and subatomic particles/ waves. Quantum mechanics is the foundation of several related disciplines, including condensed matter physics, quantum chemistry, and particle physics, and is the chief investigative tool in structural biology. Within the field of engineering, quantum mechanics plays an important part of forming the development of nanotechnology and electronics.The term "quantum mechanics" was first coined by Max Born in 1924.

The acceptance by the general physics community of quantum mechanics is due to its accurate prediction of the physical behaviour of systems, including systems where Newtonian mechanics fails. Even general relativity is limited—in ways quantum mechanics is not—in describing systems at the atomic scale or smaller, at very low or very high energies, or at the lowest temperatures. Through a century of experimentation and applied science, quantum mechanical theory has proven to be very successful and practical.

THE UNEXPECTED

Toward the end of the nineteenth century it seemed that physics had been very nearly perfected, but then paradox after paradox radically changed human concepts at fundamental levels. The models by which humans explained things on their scale in space and time failed to apply to the ultra small, the ultra massive, and to interactions at the highest speeds. A view of the universe that had seemed to be the distillation of commonplace observations was challenged by observations and theories that predicted correctly where classical physics had given impossible results. But the emerging picture was of a universe that stubbornly refused to behave the way humans had come to expect.

On the large end of the scale, relativity theory informed humans that time does not pass at the same rate for all observers, that matter can convert to energy and vice-versa, that two objects each moving at more than half the speed of light cannot have a closing speed that is equal to or greater than the speed of light, that time progresses more slowly near massive objects, etc. Things just do not work the way that experience with scales, rulers, and clocks here on earth had led humans to expect.

On the small end of the scale, wonders are even more abundant. A thing such as a photon or an electron does not have location and a traceable trajectory between the point where it is emitted and the point where it is detected. The points where such a particle may be detected are not what one would expect based on everyday experience. With a certain small probability, the detection point can even be on the other side of a solid barrier.

Probability turns out to be a highly salient factor in all interactions on this scale. The trajectory of any atomic scale object is "squishy" in the sense that any measurement intended to get clearer on the object's position forces knowledge of its velocity to become less clear, and vice-versa.In the era of classical physics, Newton and his followers believed that light consisted of particles, and others believed that light was actually a series of wave fronts propagated through some medium.

Rather than finding an experiment that would prove one or the other side correct, physicists learned that making an experiment designed to show light's frequency or other wave *characteristics* will manifest its "wave nature," and making an experiment that is designed to show its momentum or other particle characteristics will manifest its "particle nature." Furthermore, objects as large as atoms, and even some molecules, have shown their "wave nature" when addressed in the appropriate way.The most eminent of physicists have warned that if one imagines some explanation of quantum physics to "make sense," then that explanation is very likely to be flawed.

HOW THE UNEXPECTED CAME TO LIGHT

The foundations of quantum mechanics date from the early 1800s, but the real beginnings of QM date from the work of Max Planck in 1900. Albert Einstein and Niels Bohr soon made important contributions to what is now called the "old quantum theory." However, it was not until 1924 that a more complete picture emerged with Louis de Broglie's matter-wave hypothesis and the true importance of quantum mechanics became clear. Some of the most prominent scientists to subsequently contribute in the mid-1920s to what is now called the "new quantum mechanics" or "new physics" were Max Born, Paul Dirac, Werner Heisenberg, Wolfgang Pauli, and Erwin Schrödinger.

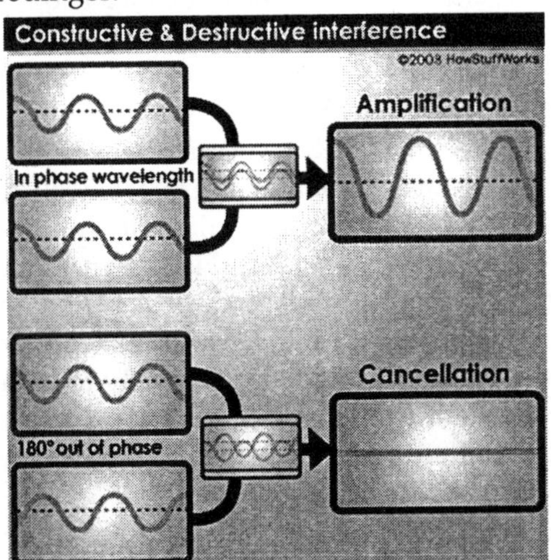

Fig. Constructive and Distructive Interfence

Later, the field was further expanded with work by Julian Schwinger, Sin-Itiro Tomonaga and Richard Feynman for the development of Quantum Electrodynamics in 1947 and by

Murray Gell-Mann in particular for the development of Quantum chromodynamics.

The interference that produces colored bands on bubbles cannot be explained by a model that depicts light as a particle. It can be explained by a model that depicts it as a wave. The drawing shows sine waves that resemble waves on the surface of water being reflected from two surfaces of a film of varying width, but that depiction of the wave nature of light is only a crude analogy.Early researchers differed in their explanations of the fundamental nature of what we now call electromagnetic radiation. Some maintained that light and other frequencies of electromagnetic radiation are composed of particles, while others asserted that electromagnetic radiation is a wave phenomenon.

In classical physics these ideas are mutually contradictory. Ever since the early days of QM, scientists have acknowledged that neither idea by itself can explain electromagnetic radiation.In 1690, Christiaan Huygens explained the laws of reflection and refraction on the basis of a wave theory. Sir Isaac Newton believed that light consisted of infinitesimally small particles which he designated "corpuscles". In 1827 Thomas Young and Augustin Fresnel made experiments on interference that showed that a corpuscular theory of light was inadequate. Then in 1873 James Clerk Maxwell showed that by making an electrical circuit oscillate it should be possible to produce electromagnetic waves. His theory made it possible to compute the speed of electromagnetic radiation purely on the basis of electrical and magnetic measurements, and the computed value corresponded very closely to the empirically measured speed of light. In 1888, Heinrich Hertz made an electrical device that actually produced what we would now call microwaves – essentially radiation at a lower frequency than visible light.

Everything up to that point suggested that Newton had been entirely wrong to regard light as corpuscular.Later experiments indicated that a packet or quantum model was needed to explain some phenomena. When light strikes an electrical conductor it causes electrons to move away from their original positions. The observed phenomenon could only be explained by assuming that the light delivers energy in definite packets. In a photoelectric device such as the light meter in a camera, light hitting the metallic detector causes electrons to move. Greater intensities of light at one frequency can cause more electrons to move, but they will not move faster. In contrast, higher frequencies of light can cause electrons to move faster.

Therefore, intensity of light controls current, but frequency of light controls voltage. These observations raised a contradiction when compared with sound waves and ocean waves, where only intensity was needed to predict the energy of the wave. In the case of light, frequency appeared to predict energy. Something was needed to explain this phenomenon and to reconcile experiments that had shown light to have particle nature with experiments that had shown it to have wave nature.

SPECTROSCOPY AND ONWARD

It is fairly easy to see a spectrum produced by white light when it passes through a prism, the bevelled edge of a mirror or a tapered pane of glass, or through drops of rain to form a rainbow. When samples which are composed purely of one single chemical element, e.g. hydrogen, are caused to emit light, they may emit light at several characteristic frequencies. The frequency profile produced is characteristic of that element. Instead of there being a wide band filled with colours from violet to red, there will be isolated bands of single colours separated by darkness. Such a display is called a line spectrum. Some lines go beyond the visible frequencies and can only be detected by special photographic film or other such devices.

Scientists hypothesized that an atom could radiate light the way the string on a fine violin radiates sound – not only with a fundamental frequency but with several higher harmonics (formed when the string divides itself into halves and other divisions that vibrate in coordination with each other as when one half of the string is going one way as the other half of the string is going the opposite way). For a long time nobody could find a mathematical way to relate the frequencies of the line spectrum of any element.

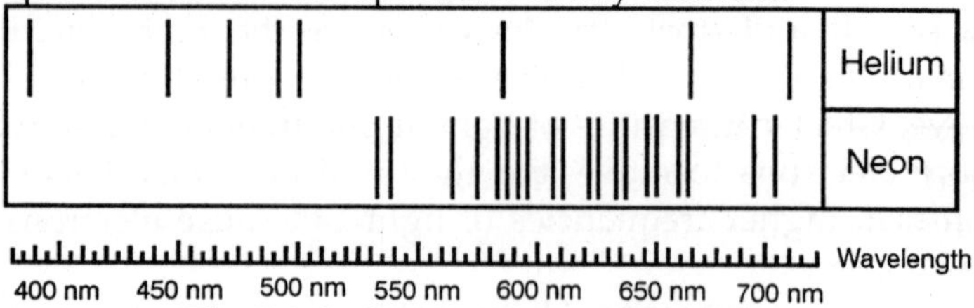

Fig. Bright-line Spectrum of Hydrogen

In 1885, Johann Jakob Balmer (1825-1898) figured out how the frequencies of atomic hydrogen are related to each other. The formula is a simple one:

$$\frac{1}{\lambda} = R\left(\frac{1}{2^2} - \frac{1}{n^2}\right)$$

where λ is wavelength, R is the Rydberg constant and n is an integer (n =3, 4,...) This formula can be generalized to apply to atoms that are more complicated than hydrogen, but we will stay with hydrogen for this general exposition. The next development was the discovery of the Zeeman effect, named after Pieter Zeeman (1865-1943).

The physical explanation of the Zeeman effect was worked out by Hendrik Antoon Lorentz (1853-1928). Lorentz hypothesized that the light emitted by hydrogen was produced by vibrating electrons.

It was possible to get feedback on what goes on within the atom because moving electrons create a magnetic field and so can be influenced by the imposition of an external magnetic field in a manner analogous to the way that one iron magnet will attract or repel another magnet.

The Zeeman effect could be interpreted to mean that light waves are originated by electrons vibrating in their orbits, but classical physics could not explain why electrons should not fall out of their orbits and into the nucleus of their atoms, nor could classical physics explain why their orbits would be such as to produce the series of discrete frequencies derived by Balmer's formula and displayed in the line spectra. Why did the electrons not produce a continuous spectrum?

OLD QUANTUM THEORY

Quantum mechanics developed from the study of electromagnetic waves through spectroscopy which includes visible light seen in the colours of the rainbow, but also other waves including the more energetic waves like ultraviolet light, x-rays, and gamma rays; and the waves with longer wavelengths including infrared waves, microwaves and radio waves. Only waves that travel at the speed of light are included in this description. Also, when the word "particle" is used below, it always refers to elementary or subatomic particles.

PLANCK'S CONSTANT

Classical physics predicted that a black-body radiator would produce infinite energy, but that result was not observed in the laboratory. If black-body radiation was dispersed into a spectrum, then the amount of energy radiated at various frequencies rose from zero at one end, peaked at a frequency related to the temperature of the radiating object, and then fell back to zero. In 1900, Max Planck developed an empirical equation that could account for the observed energy curves, but he could not harmonize it with classical theory. He concluded that the classical laws of physics do not apply on the atomic scale as had earlier been assumed.

In this theoretical account, Planck allowed all possible frequencies, all possible wavelengths. However, he restricted the energy that is delivered. "In classical physics,... the energy of a given oscillator depends merely on its amplitude, and this amplitude is subject to no restriction." But, according to Planck's theory, the energy emitted by an oscillator is strictly proportional to its frequency.

The higher the frequency, the greater the energy. To reach this theoretical conclusion, he postulated that a radiating body consisted of an enormous number of elementary oscillators, some vibrating at one frequency and some at another, with all frequencies from zero to infinity being represented. The energy E of any one oscillator was not permitted to take on any arbitrary value, but was proportional to some integral multiple of the frequency f of the oscillator. That is,

$$E = nhf.$$

where n =1, 2, 3,... The proportionality constant h is called *Planck's constant.*

One of the most direct applications is finding the energy of photons. If h is known, and the frequency of the photon is known, then the energy of the photons can be calculated. For instance, if a beam of light illuminated a target, and its frequency was 540×10^{12} hertz, then the energy of each photon would be $h \times 540 \times 10^{12}$ joules. The value of h itself is exceedingly small, about $6.6260693 \times 10^{-34}$ joule seconds. This means that the photons in the beam of light have an energy of about 3.58×10^{-19} joules or approximately 2.23 eV.

When the energy of a wave is described in this manner, it seems that the wave is carrying its energy in little packets. This discovery then seemed to remake the wave into a particle. These packets of energy carried along with the wave were called quanta by Planck. Quantum mechanics began with the discovery that energy is delivered in packets whose size is related to the frequencies of all electromagnetic waves (and to the colour of visible light since in that case frequency determines colour). Be aware, however, that these descriptions in terms of packet, wave and particle import macro-world concepts into the quantum world, where they have only provisional relevance or appropriateness.

In early research on light, there were two competing ways to describe light, either as a wave propagated through empty space, or as small particles traveling in straight lines. Because Planck showed that the energy of the wave is made up of packets, the particle analogy became favored to help understand how light delivers energy in multiples of certain set values designated as quanta of energy. Nevertheless, the wave analogy is also indispensable for helping to understand other light phenomena.

In 1905, Albert Einstein used Planck's constant to explain the photoelectric effect by postulating that the energy in a beam of light occurs in concentrations that he called light quanta, later on called photons. According to that account, a single photon of a given frequency delivers an invariant amount of energy. In other words, individual photons can deliver more or less energy, but only depending on their frequencies. Although the description that stemmed from Planck's research sounds like Newton's corpuscular account, Einstein's photon was still said to have a frequency, and the energy of the photon was accounted proportional to that frequency. The particle account had been compromised once again.

Both the idea of a wave and the idea of a particle are models derived from our everyday experience. We can only investigate their properties indirectly. We look at some phenomena, such as the rainbow of colours that we see when a thin film of oil rests on the surface of a puddle of water, and we can explain that phenomenon to ourselves by comparing light with waves. We look at other phenomena, such as the way a photoelectric meter in our camera works, and we explain it by analogy to particles colliding with the detection screen in the meter. In both cases we take concepts from our everyday experience and apply them to a world we have never seen.

Neither form of explanation, wave or particle, is entirely satisfactory. In general any model can only approximate that which it models. A model is useful only within the range of conditions where it is able to predict the real thing with accuracy. Newtonian physics is still a good predictor of many of the phenomena in our everyday life. To remind us that both "wave" and "particle" are concepts imported from our macro world to explain the world of atomic-scale phenomena, some physicists such as Banesh Hoffmann have used the term "wavicle" to refer to whatever it is that is really there.

REDUCED PLANCK'S CONSTANT

Planck's constant originally represented the energy that a light wave carries as a function of its frequency. A step in the development of this concept appeared in Bohr's work. Bohr was using a "planetary" or particle model of the electron, and could not understand why a 2π factor was essential to his experimentally derived formulae.

Later, de Broglie postulated that electrons have frequencies, just as do photons, and that the frequency of an electron must conform to the conditions for a standing wave that can exist in a certain orbit.

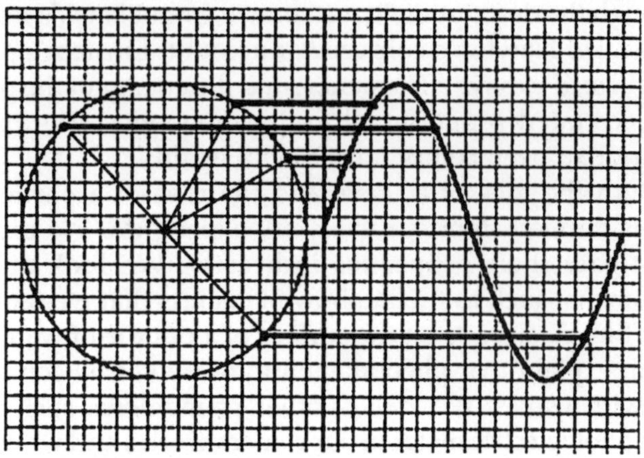

Fig. Relation between a Cycle and a Wave

That is to say, the beginning of one cycle of a wave at some point on the circumference of a circle must coincide with the end of some cycle. There can be no gap, no length along the circumference that is not participating in the vibration, and there can be no overlap of cycles. So the circumference of the orbit, C, must equal the wavelength, λ, of the electron multiplied by some positive integer (n = 1, 2, 3...). Knowing the circumference one can calculate wavelengths that fit that orbit, and knowing the radius, r, of the orbit one can calculate its circumference. To put all that in mathematical form,

$$C = 2\pi r = n\lambda$$

and so

$$\lambda = 2\pi r/n$$

and the appearance of the 2π factor is seen to occur simply because it is needed to calculate possible wavelengths (and therefore possible frequencies) when the radius of an orbit is already known.

Again in 1925 when Werner Heisenberg developed his full quantum theory, calculations involving wave analysis called Fourier series were fundamental, and so the "reduced" version of Planck's constant ($h/2\pi$) became invaluable because it includes a conversion factor to facilitate calculations involving wave analysis. Finally, when this reduced Planck's constant appeared naturally in Dirac's equation it was then given an alternate designation, "Dirac's constant." Therefore, it is appropriate to begin with an explanation of what this constant is, even though the theories that made its use convenient.

As noted above, the energy of any wave is given by its frequency multiplied by Planck's constant. A wave is made up of crests and troughs. In a wave, a cycle is defined by the return from a certain position to the same position such as from the top of one crest to the next crest. A cycle actually is mathematically related to a circle, and both have 360 degrees. A degree is a unit of measure for the amount of turn needed to produce an arc of a certain length at a given distance. A sine curve is generated by a point on the circumference of a circle as that circle rotates.

There are 2π radians per cycle in a wave, which is mathematically related to the way a circle has 360° (which are equal to two π radians). (A radian is simply the angle if a distance along the circumference of the circle is measured equal to the radius of the circle, and then lines are drawn to the centre of the circle. This forms an angle equal to 1 radian.) Since one cycle is 2π radians, when h is divided by 2π the two "2 π" factors will cancel out leaving just the radian to contend with. So, dividing h by 2π describes a constant that, when multiplied by the frequency of a wave, gives the energy in joules per radian. The reduced Planck's constant is written in mathematical formula7e as ', and is read as "h-bar".

$$\hbar = \frac{h}{2\pi}.$$

The reduced Planck's constant allows computation of the energy of a wave in units per radian instead of in units per cycle. These two constants *h* and ' are merely conversion factors between energy units and frequency units. The reduced Planck's constant is used more often than *h* (Planck's constant) alone in quantum mechanical mathematical formulas for many reasons, one of which is that angular velocity or angular frequency is ordinarily measured in radians per second so using ' that works in radians too will save a computation to put radians into degrees or vice-versa. Also, when equations relevant to those problems are written in terms of ', the frequently occurring 2π factors in numerator and denominator can cancel out, saving a computation. However, in other cases, as in the orbits of the Bohr atom, $h/2\pi$ was obtained naturally for the angular momentum of the orbits. Another expression for the relation between energy and wave length is given in electron volts for energy and angstroms for wavelength: E_{photon} (eV) = 12,400/λ(Å) — it appears not to involve h at all, but that is only because a different system of units has been used and now, numerically, the appropriate conversion factor is 12,400.

BOHR ATOM

In 1897 the particle called the electron was discovered. By means of the gold foil experiment physicists discovered that matter is, volume for volume, largely space. Once that was clear, it was hypothesized that negative charge entities called electrons surround positively charged nuclei. So at first all scientists believed that the atom must be like a miniature solar system. But that simple analogy predicted that electrons would, within about one hundredth of a microsecond, crash into the nucleus of the atom.

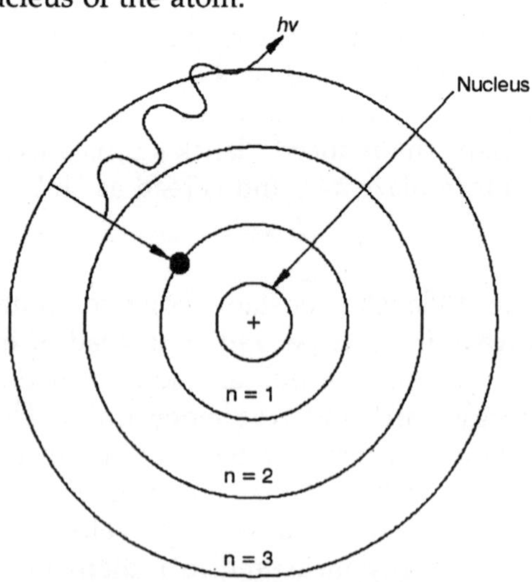

Fig. Bohr Model of the Atom

The great question of the early 20th century was, "Why do electrons normally maintain a stable orbit around the nucleus?"

In 1913, Niels Bohr removed this substantial problem by applying the idea of discrete (non-continuous) quanta to the orbits of electrons. This account became known as the Bohr model of the atom. Bohr basically theorized that electrons can only inhabit certain orbits around the atom. These orbits could be derived by looking at the spectral lines produced by atoms.

Bohr explained the orbits that electrons can take by relating the angular momentum of electrons in each "permitted" orbit to the value of h, Planck's constant. He held that an electron in the lowest orbital has a discrete angular momentum equal to $h/2\pi$. Each orbit after the initial orbit must provide for an electron's angular momentum being an integer multiple of that lowest value. He depicted electrons in atoms as being analogous to planets in a solar orbit. However, he took Planck's constant to be a fundamental quantity that introduces special requirements at this subatomic level and that explains the spacing of those "planetary" orbits.

Bohr considered one revolution in orbit to be equivalent to one cycle in an oscillator (as in Planck's initial measurements to define the constant h) which is in turn similar to one cycle in a wave. The number of revolutions per second is (or defines) what we call the frequency of that electron or that orbital. Specifying that the frequency of each orbit must be an integer multiple of Planck's constant h would only permit certain orbits, and would also fix their size.

Bohr generalized Balmer's formula for hydrogen by replacing denominator in the term 1/4 with an explicit squared variable:

$$\frac{1}{\lambda} = R_H\left(\frac{1}{m^2} - \frac{1}{n^2}\right), \quad m = 1, 2, 3, 4, 5, \ldots, \text{ and } n > m$$

where λ is the wavelength of the light, R_H is the Rydberg constant for hydrogen, and the integers n and m refer to the orbits between which electrons can transit. This generalization predicted many more line spectra than had been previously detected, and experimental confirmation of this prediction followed. It follows almost immediately that if λ is quantized as the formula above indicates, then the momentum of any photon must be quantized. The frequency of light, ν, at a given wavelengtλ is given by the relationship:

$$v = \frac{c}{\lambda} \text{ and}$$

$$\lambda = \frac{c}{\nu} \text{ and}$$

multiplying by h/h = 1,

$$\lambda = \frac{hc}{h\nu}, \text{ and we know that } E = h\text{í so}$$

which we can rewrite as:

$$\lambda = \frac{hc}{E}$$

$$\lambda = \frac{h}{E/c}, \text{ and } E/c = p \text{ (momentum) so}$$

$$\lambda = \frac{h}{p} \text{ or}$$

$$p = \frac{h}{\lambda}.$$

Beginning with line spectra, physicists were able to deduce empirically the rules according to which the orbits of electrons are determined and to discover something vital about the momentums involved that they are quantized.

Bohr next realized how the angular momentum of an electron in its orbit, L, is quantized, i.e., he determined that there is some constant value K such that when it is multiplied by Planck's constant, h, it will yield the angular momentum that pertains to the lowest orbital. When it is multiplied by successive integers it will then give the values of other possible orbitals. He later determined that K = 1/2π.

Bohr's theory represented electrons as orbiting the nucleus of an atom in a way that was amazingly different from what we see in the world of our everyday experience. He showed that when an electron changed orbits it did not move in a continuous trajectory from one orbit around the nucleus to another. Instead, it suddenly disappeared from its original orbit and reappeared in another orbit. Each distance at which an electron can orbit is a function of a quantized amount of energy.

The closer to the nucleus an electron orbits, the less energy it takes to remain in that orbital. Electrons that absorb a photon gain a quantum of energy, so they jump to an orbit that is farther from the nucleus, while electrons that emit a photon lose a quantum of energy and so jump to an inner orbital. Electrons cannot gain or lose a fractional quantum of energy, and so, it is argued, they cannot have a position that is at a fractional distance between allowed orbitals. Allowed orbitals were designated as whole numbers using the letter n with the innermost orbital being designated n = 1, the next out being n = 2, and so on. Any orbital with the same value of n is called an electron shell.

Bohr's model of the atom was essentially two-dimensional because it depicts electrons as particles in circular orbits. In this context, two-dimensional means something that can be described on the surface of a plane. One-dimensional means something that can be described by a line. Because circles can be described by their radius, which is a line segment, sometimes Bohr's model of the atom is described as one-dimensional.

WAVE-PARTICLE DUALITY

Niels Bohr determined that it is impossible to describe light adequately by the sole use of either the wave analogy or of the particle analogy. Therefore he enunciated the principle of complementarity, which is a theory of pairs, such as the pairing of wave and particle or the pairing of position and momentum. Louis de Broglie worked out the mathematical consequences of these findings. In quantum mechanics, it was found that electromagnetic waves could react in certain experiments as though they were particles and in other experiments as though they were waves. It was also discovered that subatomic particles could sometimes be described as particles and sometimes as waves. This discovery led to the theory of wave-particle duality by Louis-Victor de Broglie in 1924, which states that subatomic entities have properties of both waves and particles at the same time.

The Bohr atom model was enlarged upon with the discovery by de Broglie that the electron has wave-like properties. In accord with de Broglie's conclusions, electrons can only appear under conditions that permit a standing wave. A standing wave can be made if a string is fixed on both ends and made to vibrate. That illustration shows that the only standing waves that can occur are those with zero amplitude at the two fixed ends.

The waves created by a stringed instrument appear to oscillate in place, simply changing crest for trough in an up-and-down motion. A standing wave can only be formed when the wave's length fits the available vibrating entity. In other words, no partial fragments of wave crests or troughs are allowed. In a round vibrating medium, the wave must be a continuous formation of crests and troughs all around the circle. Each electron must be its own standing wave in its own discrete orbital.

DEVELOPMENT OF MODERN QUANTUM MECHANICS

Full Quantum Mechanical Theory

Werner Heisenberg made the full quantum mechanical theory in 1925 at the young age of 23. Following his mentor, Niels Bohr, Werner Heisenberg began to work out a theory for the quantum behaviour of electron orbitals. Because electrons could not be observed in their orbits, Heisenberg went about creating a mathematical description of quantum mechanics built on what could be observed, that is, the light emitted from atoms in their characteristic atomic spectra. Heisenberg studied the electron orbital on the model of a charged ball on a spring, an oscillator, whose motion is anharmonic. For a picture of the behaviour of a charged ball on a spring see: Vibrating Charges. Heisenberg first explained this kind of observed motion in terms of the laws of classical mechanics known to apply in the macro world, and then applied quantum restrictions, discrete (non-continuous) properties, to the picture. Doing so causes gaps to appear between the predicted orbitals so that the mathematical description he formulated would then represent only the electron orbitals predicted on the basis of the atomic spectra.

In approaching the problem that Bohr gave him to solve, Heisenberg took the strategic stance that he would not deal with unobservable quantities. He would begin formulating equations using only the quantities that could be observed.

That strategy led him to begin with the actual experimental evidence at hand: Measurements had been well established for such data as:

- The frequencies (and the mathematically related energies) emitted or absorbed by electron transitions from one of the Bohr stationary orbits, known to be associated with the bright line spectra.
- The "transition amplitude" or likelihood of transition from any given orbit to any given orbit, known from the strength of the various lines in the bright spectrum, etc.

From classical formulas that would characterize those phenomena Heisenberg created analogous formulas that took account of quantum conditions. Formulas that followed from the fundamental decisions made at this point resulted in good results but results that were sometimes not what one might expect.

In the paper wherein he introduced quantum mechanics to the world he cautions, "A significant difficulty arises, however, if we consider two quantities $x(t)$, $y(t)$, and ask after their product....Whereas in classical $x(t)y(t)$ is always equal to $y(t)x(t)$, this is not necessarily the case in quantum theory." When the predicted values are exhibited in matrix form and multiplications are performed, the nature of the difficulty appears in a form that is more familiar to mathematicians. More significantly, empirical studies validate the theoretical results and suggest that there is something of deep importance in that the difference between x(t)y(t) and y(t)x(t) is a value related to Planck's constant.

Schema for a table of transition frequencies (produced when electrons change orbitals):

Electron States	*S1*	*S2*	*S3*	*S4*	*S5*	....
S1	f1→1	f2→1	f3→1	f4→1	f5→1	.....
S2	f1→2	f2→2	f3→2	f4→2	f5→2	.....
S3	f1→3	f2→3	f3→3	f4→3	f5→3	.....
S4	f1→4	f2→4	f3→4	f4→4	f5→4	.....
S5	f1→5	f2→5	f3→5	f4→5	f5→5	.....
S.....	.....	.....	.....	.....	.....	.....

Schema for a related table showing the transition amplitudes:

Electron States	*S1*	*S2*	*S3*	*S4*	*S5*	....
S1	a1′!1	a2′!1	a3′!1	a4′!1	a5′!1	.....
S2	a1′!2	a2′!2	a3′!2	a4′!2	a5′!2	.....
S3	a1′!3	a2′!3	a3′!3	a4′!3	a5′!3	.....
S4	a1′!4	a2′!4	a3′!4	a4′!4	a5′!4	.....
S5	a1′!5	a2′!5	a3′!5	a4′!5	a5′!5	.....
S.....	.....	.....	.....	.....	.....	.....

As related above, Heisenberg developed ways of meaningfully relating the information in tables such as these in a mathematical way. Empirically filling in the values for tables involving quantum quantities is not a simple procedure since any measurement made on a single system gives that one value but has the potential of changing other values. So large numbers of identical copies of the system in question must be prepared, and a single measure made on each system. Multiple experiments to determine the same characteristics are made, and the results are averaged. Even then, precise measurements of all characteristics of the system as they would appear simultaneously cannot be provided because of quantum uncertainty.

A precise determination of one characteristic's value necessarily creates an uncertainty in the value of its correlate. "Certain pairs of observables simply *cannot* be simultaneously measured to an arbitrarily high level of precision." If simultaneous measurements are made of correlated characteristics in multiple identical systems, there will inevitably be differences in the measurements such that the difference between their products is equal to or greater than $\hbar/2$."

In 1925 Heisenberg published a paper entitled "Quantum-mechanical re-interpretation of kinematic and mechanical relations" relating his discoveries. So ended the old quantum theory and began the age of quantum mechanics. Heisenberg's paper gave few details that might aid readers in determining how he actually contrived to get his results for the one-dimensional models he used to form the hypothesis that proved so useful. In his paper, Heisenberg proposed to "discard all hope of observing hitherto unobservable quantities, such as the position and period of the electron", and restrict himself strictly to actually observable quantities.

He needed mathematical rules for predicting the relations actually observed in nature, and the rules he produced worked differently depending on the sequence in which they were applied. "It quickly became clear that the *non-commutativity* of kinematical quantities in quantum theory was the really essential new technical idea in the paper."

The special type of multiplication that turned out to be required in his formula was most elegantly described by using special arrays of numbers called matrices. In ordinary situations it does not matter in which order the operations involved in multiplication are performed, but matrix multiplication does not commute. Essentially that means that it matters which order given operations are performed in. Multiplying matrix A by matrix B is not the same as multiplying matrix B by matrix A.

In symbols, A×B is in general not equal to B×A. (The important thing in quantum theory is that it turned out to matter whether one experimentally measures velocity first and then immediately measures position, or vice-versa.) The matrix convention turned out to be a convenient way of organizing information and making clear the exact sequence in which calculations must be made, and it reflects in a symbolic form the unexpected results obtained in the real world.

Heisenberg approached quantum mechanics from the historical perspective that treated an electron as an oscillating charged particle. Bohr's use of this analogy had already allowed him to explain why the radii of the orbits of electrons could only take on certain values. It followed from this interpretation of the experimental results available and the quantum theory that Heisenberg subsequently created that an electron could not be at any intermediate position between two "permitted" orbits. Therefore electrons were described as "jumping" from orbit to orbit. The idea that an electron might now be in one place and an instant later be in some other place without having traveled between the two points was one of the earliest indications of the "spookiness" of quantum phenomena. Although the scale is smaller, the "jump" from orbit to orbit is as strange and unexpected as would be a case in which someone stepped out of a doorway in London onto the streets of Los Angeles. Quantum tunneling is one instance in which electrons seem to be able to move in the "spooky" way that Heisenberg ascribed to their actions within atoms.

Amplitudes of position and momentum that have a period of 2π like a cycle in a wave are called Fourier series variables. Heisenberg described the particle-like properties of the electron in a wave as having position and momentum in his matrix mechanics. When these amplitudes of position and momentum are measured and multiplied together, they give intensity. However, he found that when the position and momentum were multiplied together in that respective order, and then the momentum and position were multiplied together in that respective order, there was a difference or deviation in intensity between them of $h/2\pi$. Heisenberg would not understand the reason for this deviation until two more years had passed, but for the time being he satisfied himself with the idea that the math worked and provided an exact description of the quantum behaviour of the electron.

Matrix mechanics was the first complete definition of quantum mechanics, its laws, and properties that described fully the behaviour of the electron. It was later extended to apply to all subatomic particles.

Very soon after matrix mechanics was introduced to the world, Schrödinger, acting independently, produced a quantum wave theory that appeared to have no similarities whatsoever to Heisenberg's theory. It was computationally easier and avoided some of the odd-sounding ideas like "quantum leaps" of an electron from one orbit to another. But within a short time Schrödinger himself had shown that the two theories produced essentially the same results in all situations. Finally, Dirac made the idea of non-commutativity central to his own theory and proved the formulations of Heisenberg and of Schrödinger to be special cases of his own theory.

Schrodinger Wave Equation

Because particles could be described as waves, later in 1925 Erwin Schrödinger analyzed what an electron would look like as a wave around the nucleus of the atom. Using this model, he formulated his equation for particle waves. Rather than explaining the atom by analogy to satellites in planetary orbits, he treated everything as waves whereby each electron has its own unique wavefunction. A wavefunction is described in Schrödinger's equation by three properties. The three properties were:

- An "orbital" designation, indicating whether the particle wave is one that is closer to the nucleus with less energy or one that is farther from the nucleus with more energy,
- The shape of the orbital, i.e., an indication that orbitals were not just spherical but other shapes, and
- The inclination of the orbital, which determines the magnetic moment of the orbital around the z-axis.

These three properties were called collectively the wavefunction of the electron and are said to describe the quantum state of the electron. "Quantum state" means the collective properties of the electron describing what we can say about its condition at a given time. For the electron, the quantum state is described by its wavefunction, which is designated in physics by the Greek letter ø (psi, pronounced "sigh").

The three properties of Schrödinger's equation that describe the wavefunction of the electron and, therefore, also describe the quantum state of the electron as described in the previous paragraph are each called quantum numbers. The first property that describes the orbital was numbered according to Bohr's model where n is the letter used to describe the energy of each orbital. This is called the principal quantum number.

The next quantum number that describes the shape of the orbital is called the azimuthal quantum number and it is represented by the letter l (lower case L). The shape is caused by the angular momentum of the orbital. The rate of change of the angular momentum of any system is equal to the resultant external torque acting on that system. In other words, angular momentum represents the resistance of a spinning object to speed up or slow down under the influence of external force.

The azimuthal quantum number "l" represents the orbital angular momentum of an electron around its nucleus. However, the shape of each orbital has its own letter as well. So for the letter "l" there are other letters to describe the shapes of "l". The first shape is spherical and is described by the letter s. The next shape is like a dumbbell and is described by the letter p. The other shapes of orbitals become more complicated and are described by the letters d, f, and g. For the shape of a carbon atom. The third quantum number of Schrödinger's equation describes the magnetic moment of the electron and is designated by the letter m and sometimes as the letter m with a subscript l because the magnetic moment depends upon the second quantum number l.

In May 1926 Schrödinger published a proof that Heisenberg's matrix mechanics and his own wave mechanics gave equivalent results: mathematically they were the same theory. Yet both men disagreed on the interpretation of this theory. Heisenberg saw no problem in the existence of discontinuous quantum jumps, while Schrödinger hoped that a theory based on continuous wave-like properties could avoid this "nonsense about quantum jumps".

Uncertainty Principle

In 1927, Heisenberg made a new discovery on the basis of his quantum theory that had further practical consequences of this new way of looking at matter and energy on the atomic scale. In Heisenberg's matrix mechanics formula, he encountered an error or difference of $h/2\pi$ between position and momentum.

The more certain the position of a particle is determined, the less certain the momentum is known, and $h/2\pi$ is the lower limit of the uncertainty involved. This conclusion came to be called "Heisenberg's Indeterminacy Principle", or Heisenberg's Uncertainty Principle.For moving particles in quantum mechanics, there is simply a certain degree of exactness and precision that is missing. The observer can be precise when taking a measurement of position or can be precise when taking a measurement of momentum, but there is an inverse imprecision when measuring both at the same time as in the case of a moving particle like the electron. One consequence of the uncertainty principle was that the electron could no longer be considered as in an exact location in its orbital.

Rather the electron had to be described by every point where the electron could possibly inhabit. Calculating points of probable location for the electron in its known orbital created the picture of a cloud of points in a spherical shape for the orbital of a hydrogen atom which points gradually faded out nearer to the nucleus and farther from the nucleus. This picture may be termed a probability distribution. Therefore, the Bohr atom number n for each orbital became known as an *n-sphere* in the three dimensional atom and was pictured as a probability cloud where the electron surrounded the atom all at once.

This led to the further description by Heisenberg that if a measurement of the electron was not being taken that it could not be described in one particular location but was everywhere in the electron cloud at once. In other words, quantum mechanics cannot give exact results, but only the probabilities for the occurrence of a variety of possible results.

Heisenberg went further and said that the path of a moving particle only comes into existence once we observe it. However strange and counter-intuitive this assertion may seem, quantum mechanics does still tell us the location of the electron's orbital, its probability cloud. Heisenberg was speaking of the particle itself, not its orbital which is in a known probability distribution.

It is important to note that although Heisenberg used infinite sets of positions for the electron in his matrices, this does not mean that the electron could be anywhere in the universe. Rather there are several laws that show the electron must be in one localized probability distribution. An electron is described by its energy in Bohr's atom which was carried over to matrix mechanics.

Therefore, an electron in a certain n-sphere had to be within a certain range from the nucleus depending upon its energy. This restricts its location. Also, the number of places an electron can be is also called "the number of cells in its phase space". The Uncertainty Principle set a lower limit to how finely one can chop up classical phase space, so the number of places that an electron can be in its orbital becomes finite. An electron's location in an atom is defined to be in its orbital, but stops at the nucleus and before the next n-sphere orbital begins.

Classical physics had shown since Newton that if the position of stars and planets and details about their motions were known then where they will be in the future can be predicted.

For subatomic particles, Heisenberg denied this notion showing that due to the uncertainty principle one cannot know the precise position and momentum of a particle at a given instant, so its future motion cannot be determined, but only a range of possibilities for the future motion of the particle can be described.

These notions arising from the uncertainty principle only arise at the subatomic level and were a consequence of wave-particle duality. As counter-intuitive as they may seem, quantum mechanical theory with its uncertainty principle has been responsible for major improvements in the world's technology from computer components to fluorescent lights to brain scanning techniques.

WAVEFUNCTION COLLAPSE

Schrödinger's wave equation with its unique wavefunction for a single electron is also spread out in a probability distribution like Heisenberg's quantized particle-like electron. This is because a wave is naturally a widespread disturbance and not a point particle.

Therefore, Schrödinger's wave equation has the same predictions made by the uncertainty principle because uncertainty of location is built into the definition of a widespread disturbance like a wave.

Uncertainty only needed to be defined from Heisenberg's matrix mechanics because the treatment was from the particle-like aspects of the electron.

Schrödinger's wave equation shows that the electron is in the probability cloud at all times in its probability distribution as a wave that is spread out. Max Born discovered in 1928 that when the square of Schrödinger's wavefunction (psi-squared) is computed the electron's location as a probability distribution is obtained.

Therefore, to the extent that a measurement of the position of an electron can be made at an exact location instead of as a probability distribution, the electron appears to momentarily cease to have wave-like properties. Without wave-like properties, none of Schrödinger's definitions of the electron being wave-like makes sense. The measurement of the position of the particle nullifies the simple wave-like properties and the one body form of Schrödinger's equation then fails.

Because the electron can no longer be described by its separate wavefunction when measured, due to its wave length becoming much shorter and its becoming entangled with the particles of the measuring apparatus, this is called wavefunction collapse.

Eigenstates and Eigenvalues

The term eigenstate is derived from the German/Dutch word *"eigen,"* which means "inherent" or "characteristic." The word eigenstate is descriptive of the measured state of some object that possesses quantifiable characteristics such as position, momentum, etc. The state being measured and described must be an "observable" (i.e. something that can be experimentally measured either directly or indirectly like position or momentum), and must have a definite value. In the everyday world, it is natural and intuitive to think of everything being in its own eigenstate.

Everything appears to have a definite position, a definite momentum, a definite value of measure, and a definite time of occurrence. However, quantum mechanics affirms that it is impossible to pinpoint exact values for the momentum of a certain particle like an electron in a given location at a particular moment in time, or, alternatively, that it is impossible to give an exact location for such an object when the momentum has been measured. Due to the uncertainty principle, statements regarding both the position and momentum of particles can only be given in terms of a range of probabilities, a "probability distribution". Eliminating uncertainty in one term maximizes uncertainty in regard to the second parameter.

Therefore it became necessary to have a way to clearly formulate the difference between the state of something that is uncertain in the way just described, such as an electron in a probability cloud, and effectively contrast it to the state of something that is not uncertain, something that has a definite value. When something is in the condition of being definitely "pinned-down" in some regard, it is said to possess an eigenstate. For example, if the position of an electron has been made definite, it is said to have an eigenstate of position.

A definite value, such as the position of an electron that has been successfully located, is called the eigenvalue of the eigenstate of position. The German word *"eigen"* was first used in this context by the mathematician David Hilbert in 1904. Schrödinger's wave equation gives wavefunction solutions, meaning the possibilities where the electron might be, just as does Heisenberg's probability distribution. As stated above, when a wavefunction collapse occurs because something has been done to locate the position of an electron, the electron's state becomes an eigenstate of position, meaning that the position has a known value.

THE PAULI EXCLUSION PRINCIPLE

The Pauli Exclusion Principle states that no electron (or other fermion) can be in the same quantum state as another within an atom.

Wolfgang Pauli developed the Exclusion Principle from what he called a "two-valued quantum degree of freedom" to account for the observation of a doublet, meaning a pair of lines differing by a small amount (e.g., something on the order of 0.15Å), in the spectrum of the hydrogen atom. The observation meant that there was more energy in the electron orbital from magnetic moment than had previously been described. In early 1925, the young physicists Uhlenbeck and Goudsmit introduced a theory that the electron rotates in space in the same way that the earth rotates on its axis.

This would account for the missing magnetic moment and allow for two electrons in the same orbital to be different if their spins were in opposite directions, thus satisfying the Exclusion Principle. A new quantum number was then needed, one to represent the momentum involved in the rotation of each electron.

One concise statement of Pauli's principle is as follows:

There cannot exist an atom in such a quantum state that two electrons within have the same set of quantum numbers.

An electron has four quantum numbers:

- n, the total quantum number of the orbit
- l, the azimuthal quantum number
- m_l, the orbital magnetic quantum number
- m_s, the spin magnetic quantum number

"In the case of a helium atom with two electrons in the 1*s* orbital, the Pauli exclusion principle requires that the two electrons differ in the value of one quantum number. Their values of n, l, and m_l are the same; moreover, they have the same spin quantum number, s = 1/2. Accordingly they must differ in the value of m_s, which can have the value of +½ for one electron and –½ for the other."

Dirac wave Equation

In 1928, Paul Dirac extended the Pauli equation, which described spinning electrons, to account for relativity. By using the simplest electromagnetic interaction, he was able to predict the value of the magnetic moment associated with the electron's spin, and found the experimentally observed value, a value too large to be that of a spinning classical charged sphere. He was able to solve for the spectral lines of the Hydrogen atom, reproducing Sommerfeld's successful formula for the fine-structure splitting of the spectral lines from sound physical principles.

Dirac's theory suffered from a problem of negative energy, which required Dirac to introduce antiparticles and a dynamical vacuum and led to many-particle quantum field theory. All of the above development of quantum theory was based mainly on the atomic spectrum of the hydrogen atom. This is due to the fact that each atom of each element produces a unique pattern of spectral lines when light from each different kind of element is passed through a prism. Scientists could not study the electron and nucleus of the atom itself because they cannot be seen.

Even today with High-resolution Scanning Tunneling Electron Microscopes we can only get images of the atom as a blurry fuzzball. However, the spectral lines of the atom reveal the orbits of electrons and the energy that can be expected. It was basically this study of the spectroscopic analysis of first the hydrogen atom and then the helium atom that led to quantum theory.

Therefore, the mathematical formulas were made to fit the picture of the atomic spectrum. That is why quantum mechanics is sometimes referred to as a form of mathematical physics.

Quantum Entanglement

Albert Einstein rejected Heisenberg's Uncertainty Principle insofar as it seemed to imply more than a necessary limitation on human ability to actually know what occurs in the quantum realm. In a letter to Max Born in 1926, Einstein famously declared that "God does not play dice". The bare surface level prescription for making predictions from quantum mechanics, based on Born's rule for computing probabilities, became known as the Copenhagen Interpretation of quantum mechanics.

Bohr spent many years developing and refining this interpretation in light of Einstein's objections. After the 1930 Solvay conference, Einstein never again challenged the Copenhagen interpretation on technical grounds, but did not cease a philosophical attack on the interpretation, on the grounds of realism and locality (i.e., the idea that there is no action at a distance).

Einstein, in trying to show that quantum theory was not a complete theory, recognized that the theory predicted that two or more particles that had interacted in the past can exhibit strong correlations later on when various measurements are made on them.

He wanted this seeming interaction to be explained in a classical way through their common past, and preferably not by a "spooky action at a distance". The argument is worked out in a famous paper devoted to what is now called the EPR paradox. Assuming what is now usually called "local realism", the EPR attempts to show from quantum theory that particles simultaneously possess both position and momentum, while according to the Copenhagen interpretation, only one of these two properties actually exists, and only in the moment that it is being measured.

Einstein considered this conclusion a proof that quantum theory is incomplete in that it refuses to physical properties which objectively exist in nature.

The feature of quantum theory leading to these paradoxes is called quantum entanglement. Quantum entanglement refers to situations wherein the properties of several discrete objects cannot be described simply by considering them separately, even after taking account of the history of their past interaction. A quantum interaction can affect the entangled quantum entity in a like manner even though the two of them are widely separated. Bohr's original response to Einstein was that entangled particles are part of one indivisible system.

Einstein's challenge led to decades of substantial research into quantum entanglement. Research would seem to confirm Bohr's assertion that two or more entangled particles must be viewed together as one whole, and, moreover, that difficulties only arise by insisting on the reality of outcomes of measurements that are not made anyway. It might be said that God does indeed throw dice, though rather peculiar ones. A real dice throw can be completely understood using the laws of classical mechanics, and the outcome is merely a function of the initial conditions. However the outcome of tossing quantum "dice" has no antecedent; it has as yet no discernible cause.

INTERPRETATIONS

Despite the success of quantum mechanics, it does have some controversial elements. For example, the behaviour of microscopic objects described in quantum mechanics is very different from our everyday experience, which may provoke some degree of incredulity. Most of classical physics is now recognized to be composed of special cases of quantum physics theory and/or relativity theory. Dirac brought relativity theory to bear on quantum physics so that it could properly deal with events that occur at a substantial fraction of the speed of light. Classical physics, however, also deals with mass attraction (gravity), and no one has yet been able to bring gravity into a unified theory with the relativized quantum theory.Also interesting, according to the correspondence principle and Ehrenfest's theorem as a system becomes larger or more massive (action >> h) the classical dynamics tends to emerge (with some exceptions such as superfluidity).

This is why we can usually ignore quantum mechanics when dealing with everyday objects; instead the classical description will suffice. Even so, trying to make sense of quantum theory is an ongoing process that has spawned a number of interpretations of quantum theory, ranging from the conventional Copenhagen Interpretation to hidden variables and many worlds interpretations. There seems to be no end in sight to the philosophical musings on the subject; however the empirical or technical success of the theory is unrivaled; all modern fundamental physical theories are quantum theories, relativity being subsumed within quantum field theories.

RELATIVITY AND QUANTUM MECHANICS

The modern world of physics is founded on two tested and demonstrably sound theories of general relativity and quantum mechanics —theories which appear to contradict one another. The defining postulates of both Einstein's theory of relativity and quantum theory are indisputably supported by rigorous and repeated empirical evidence. However, while they do not directly contradict each other theoretically they are resistant to being incorporated within one cohesive model.Einstein himself is well known for rejecting some of the claims of quantum mechanics. While clearly inventive in this field, he did not accept the more philosophical consequences and interpretations of quantum mechanics, such as the lack of deterministic causality and the assertion that a single subatomic particle can occupy numerous areas of space at one time.

He also was the first to notice some of the apparently exotic consequences of entanglement and used them to formulate the Einstein-Podolsky-Rosen paradox, in the hope of showing that quantum mechanics has unacceptable implications. This was 1935, but in 1964 it was shown by John Bell that Einstein's assumption that quantum mechanics is correct, but has to be completed by *hidden variables*, was based on wrong philosophical assumptions: according to the paper of J. Bell and the Copenhagen interpretation (the common interpretation of quantum mechanics by physicists for decades), and contrary to Einstein's ideas, quantum mechanics is

- Neither a "realistic" theory (since quantum measurements do not *state* pre-existing properties, but rather they *prepare* properties)
- Nor a *local* theory (essentially not, because the state vector $|\psi\rangle$ determines simultaneously the probability amplitudes at all sites, $|\psi\rangle \rightarrow \psi(r), \forall r$).

The Einstein-Podolsky-Rosen paradox shows in any case that there exist experiments by which one can measure the state of one particle and instantaneously change the state of its entangled partner, although the two particles can be an arbitrary distance apart; however, this effect does not violate causality, since no transfer of information happens. These experiments are the basis of some of the most topical applications of the theory, quantum cryptography, which works well, although at small distances of typically $\leq$ 1000 km, being on the market since 2004.

There do exist quantum theories which incorporate *special* relativity-for example, quantum electrodynamics (QED), which is currently the most accurately tested physical theory-and these lie at the very heart of modern particle physics. Gravity is negligible in many areas of particle physics, so that unification between general relativity and quantum mechanics is not an urgent issue in those applications. However, the lack of a correct theory of quantum gravity is an important issue in cosmology.

ATTEMPTS AT A UNIFIED THEORY

Inconsistencies arise when one tries to join the quantum laws with general relativity, a more elaborate description of spacetime which incorporates gravitation. Resolving these inconsistencies has been a major goal of twentieth- and twenty-first-century physics. Many prominent physicists, including Stephen Hawking, have labored in the attempt to discover a "Grand Unification Theory" that combines not only different models of subatomic physics, but also derives the universe's four forces—the strong force, electromagnetism, weak force, and gravity— from a single force or phenomenon.

QUANTUM MECHANICS AND CLASSICAL PHYSICS

Predictions of quantum mechanics have been verified experimentally to a very high degree of accuracy. Thus, the current logic of correspondence principle between classical and quantum mechanics is that all objects obey laws of quantum mechanics, and classical mechanics is just a quantum mechanics of large systems (or a statistical quantum mechanics of a large collection of particles).

Laws of classical mechanics thus follow from laws of quantum mechanics at the limit of large systems or large quantum numbers.Main differences between classical and quantum theories have already been mentioned above in the remarks on the Einstein-Podolsky-Rosen paradox. Essentially the difference boils down to the statement that quantum mechanics is coherent (addition of *amplitudes*), whereas classical theories are incoherent (addition of *intensities*). Thus, such quantities as *coherence lengths* and *coherence times* come into play.

For microscopic bodies the extension of the system is certainly much smaller than the coherence length; for macroscopic bodies one expects that it should be the other way round.Many "macroscopic" properties of "classic" systems are direct consequences of quantum behaviour of its parts. For example, stability of bulk matter (which consists of atoms and molecules which would quickly collapse under electric forces alone), rigidity of this matter, mechanical, thermal, chemical, optical and magnetic properties of this matter—they are all results of interaction of electric charges under the rules of quantum mechanics.

Whilst seemingly exotic behaviour of matter posited by quantum mechanics and relativity theory become more apparent when dealing with extremely fast-moving or extremely tiny particles, the laws of classical "Newtonian" physics still remain accurate in predicting the behaviour of surrounding ("large") objects—of the order of the size of large molecules and bigger—at velocities much smaller than the velocity of light.

There are numerous mathematically equivalent formulations of quantum mechanics. One of the oldest and most commonly used formulations is the transformation theory proposed by Cambridge theoretical physicist Paul Dirac, which unifies and generalizes the two earliest formulations of quantum mechanics, matrix mechanics (invented by Werner Heisenberg) and wave mechanics.

In this formulation, the instantaneous state of a quantum system encodes the probabilities of its measurable properties, or "observables". Examples of observables include energy, position, momentum, and angular momentum. Observables can be either continuous (e.g., the position of a particle) or discrete (e.g., the energy of an electron bound to a hydrogen atom). Generally, quantum mechanics does not assign definite values to observables.

Instead, it makes predictions about probability distributions; that is, the probability of obtaining each of the possible outcomes from measuring an observable. Naturally, these probabilities will depend on the quantum state at the instant of the measurement.

There are, however, certain states that are associated with a definite value of a particular observable. These are known as "eigenstates" of the observable ("eigen" can be roughly translated from German as inherent or as a characteristic). In the everyday world, it is natural and intuitive to think of everything being in an eigenstate of every observable. Everything appears to have a definite position, a definite momentum, and a definite time of occurrence.

However, quantum mechanics does not pinpoint the exact values for the position or momentum of a certain particle in a given space in a finite time; rather, it only provides a range of probabilities of where that particle might be. Therefore, it became necessary to use different words.for the state of something having an *uncertainty relation* and a state that has a *definite* value. The latter is called the "eigenstate" of the property being measured.

For example, consider a free particle. In quantum mechanics, there is wave-particle duality so the properties of the particle can be described as a wave. Therefore, its quantum state can be represented as a wave, of arbitrary shape and extending over all of space, called a wave function. The position and momentum of the particle are observables. The Uncertainty Principle of quantum mechanics states that both the position and the momentum cannot simultaneously be known with infinite precision at the same time.

However, one can measure just the position alone of a moving free particle creating an eigenstate of position with a wavefunction that is very large at a particular position x, and almost zero everywhere else. If one performs a position measurement on such a wavefunction, the result x will be obtained with almost 100% probability. In other words, the position of the free particle will almost be known.

This is called an eigenstate of position (mathematically more precise: a *generalized eigenstate (eigendistribution)*). If the particle is in an eigenstate of position then its momentum is completely unknown. An eigenstate of momentum, on the other hand, has the form of a plane wave. It can be shown that the wavelength is equal to h/p, where h is Planck's constant and p is the momentum of the eigenstate. If the particle is in an eigenstate of momentum then its position is completely blurred out.

Usually, a system will not be in an eigenstate of whatever observable we are interested in. However, if one measures the observable, the wavefunction will instantaneously be an eigenstate of that observable. This process is known as wavefunction collapse. It involves expanding the system under study to include the measurement device, so that a detailed quantum calculation would no longer be feasible and a classical description must be used. If one knows the corresponding wave function at the instant before the measurement, one will be able to compute the probability of collapsing into each of the possible eigenstates.

For example, the free particle in the previous example will usually have a wavefunction that is a wave packet centered around some mean position x_0, neither an eigenstate of position nor of momentum. When one measures the position of the particle, it is impossible to predict with certainty the result that we will obtain. It is probable, but not certain, that it will be near x_0, where the amplitude of the wave function is large. After the measurement is performed, having obtained some result x, the wave function collapses into a position eigenstate centered at x.

Wave functions can change as time progresses. An equation known as the Schrödinger equation describes how wave functions change in time, a role similar to Newton's second law in classical mechanics. The Schrödinger equation, applied to the aforementioned example of the free particle, predicts that the centre of a wave packet will move through space at a constant velocity, like a classical particle with no forces acting on it.

However, the wave packet will also spread out as time progresses, which means that the position becomes more uncertain. This also has the effect of turning position eigenstates into broadened wave packets that are no longer position eigenstates.Some wave functions produce probability distributions that are constant in time. Many systems that are treated dynamically in classical mechanics are described by such "static" wave functions.

For example, a single electron in an unexcited atom is pictured classically as a particle moving in a circular trajectory around the atomic nucleus, whereas in quantum mechanics it is described by a static, spherically symmetric wavefunction surrounding the nucleus. The time evolution of wave functions is deterministic in the sense that, given a wavefunction at an initial time, it makes a definite prediction of what the wavefunction will be at any later time.

During a measurement, the change of the wavefunction into another one is not deterministic, but rather unpredictable, i.e., random.The probabilistic nature of quantum mechanics thus stems from the act of measurement. This is one of the most difficult aspects of quantum systems to understand. It was the central topic in the famous Bohr-Einstein debates, in which the two scientists attempted to clarify these fundamental principles by way of thought experiments. In the decades after the formulation of quantum mechanics, the question of what constitutes a "measurement" has been extensively studied. Interpretations of quantum mechanics have been formulated to do away with the concept of "wavefunction collapse" the relative state interpretation. The basic idea is that when a quantum system interacts with a measuring apparatus, their respective wavefunctions become entangled, so that the original quantum system ceases to exist as an independent entity.

Mathematical Formulation

In the mathematically rigorous formulation of quantum mechanics, developed by Paul Dirac and John von Neumann, the possible states of a quantum mechanical system are represented by unit vectors (called "state vectors") residing in a complex separable Hilbert space well defined up to a complex number of norm 1 (the phase factor). In other words, the possible states are points in the projectivization of a Hilbert space, usually called the complex projective space. The exact nature of this Hilbert space is dependent on the system; for example, the state space for position and momentum states is the space of square-integrable functions, while the state space for the spin of a single proton is just the product of two complex planes.

Each observable is represented by a maximally-Hermitian linear operator acting on the state space. Each eigenstate of an observable corresponds to an eigenvector of the operator, and the associated eigenvalue corresponds to the value of the observable in that eigenstate. If the operator's spectrum is discrete, the observable can only attain those discrete eigenvalues.The time evolution of a quantum state is described by the Schrödinger equation, in which the Hamiltonian, the operator corresponding to the total energy of the system, generates time evolution.

The inner product between two state vectors is a complex number known as a probability amplitude. During a measurement, the probability that a system collapses from a given initial state to a particular eigenstate is given by the square of the absolute value of the probability amplitudes between the initial and final states. The possible results of a measurement are the eigenvalues of the operator - which explains the choice of *Hermitian* operators, for which all the eigenvalues are real.

We can find the probability distribution of an observable in a given state by computing the spectral decomposition of the corresponding operator. Heisenberg's uncertainty principle is represented by the statement that the operators corresponding to certain observables do not commute.The Schrödinger equation acts on the entire probability amplitude, not merely its absolute value.

Whereas the absolute value of the probability amplitude encodes information about probabilities, its phase encodes information about the interference between quantum states. This gives rise to the wave-like behaviour of quantum states.

It turns out that analytic solutions of Schrödinger's equation are only available for a small number of model Hamiltonians, of which the quantum harmonic oscillator, the particle in a box, the hydrogen molecular ion and the hydrogen atom are the most important representatives.

Even the helium atom, which contains just one more electron than hydrogen, defies all attempts at a fully analytic treatment. There exist several techniques for generating approximate solutions. For instance, in the method known as perturbation theory one uses the analytic results for a simple quantum mechanical model to generate results for a more complicated model related to the simple model by, for example, the addition of a weak potential energy.

Another method is the "semi-classical equation of motion" approach, which applies to systems for which quantum mechanics produces weak deviations from classical behaviour. The deviations can be calculated based on the classical motion. This approach is important for the field of quantum chaos.An alternative formulation of quantum mechanics is Feynman's path integral formulation, in which a quantum-mechanical amplitude is considered as a sum over histories between initial and final states; this is the quantum-mechanical counterpart of action principles in classical mechanics.

Interactions with Other Scientific Theories

The fundamental rules of quantum mechanics are very broad. They assert that the state space of a system is a Hilbert space and the observables are Hermitian operators acting on that space, but do not tell us which Hilbert space or which operators, or if it even exists. These must be chosen appropriately in order to obtain a quantitative description of a quantum system. An important guide for making these choices is the correspondence principle, which states that the predictions of quantum mechanics reduce to those of classical physics when a system moves to higher energies or equivalently, larger quantum numbers.

In other words, classic mechanics is simply a quantum mechanics of large systems. This "high energy" limit is known as the *classical* or *correspondence limit*. One can therefore start from an established classical model of a particular system, and attempt to guess the underlying quantum model that gives rise to the classical model in the correspondence limit.

When quantum mechanics was originally formulated, it was applied to models whose correspondence limit was non-relativistic classical mechanics. For instance, the well-known model of the quantum harmonic oscillator uses an explicitly non-relativistic expression for the kinetic energy of the oscillator, and is thus a quantum version of the classical harmonic oscillator.Early attempts to merge quantum mechanics with special relativity involved the replacement of the Schrödinger equation with a covariant equation such as the Klein-Gordon equation or the Dirac equation. While these theories were successful in explaining many experimental results, they had certain unsatisfactory qualities stemming from their neglect of the relativistic creation and annihilation of particles.

A fully relativistic quantum theory required the development of quantum field theory, which applies quantization to a field rather than a fixed set of particles. The first complete quantum field theory, quantum electrodynamics, provides a fully quantum description of the electromagnetic interaction.

The full apparatus of quantum field theory is often unnecessary for describing electrodynamic systems. A simpler approach, one employed since the inception of quantum mechanics, is to treat charged particles as quantum mechanical objects being acted on by a classical electromagnetic field. For example, the elementary quantum model of the hydrogen atom describes the electric field of the hydrogen atom using a classical $-\frac{e^2}{4\pi \epsilon_0} \frac{1}{r}$ Coulomb potential. This "semi-classical" approach fails if quantum fluctuations in the electromagnetic field play an important role, such as in the emission of photons by charged particles.

Quantum field theories for the strong nuclear force and the weak nuclear force have been developed.

The quantum field theory of the strong nuclear force is called quantum chromodynamics, and describes the interactions of the subnuclear particles: quarks and gluons.

The weak nuclear force and the electromagnetic force were unified, in their quantized forms, into a single quantum field theory known as electroweak theory, by the physicists Abdus Salam, Sheldon Glashow and Steven Weinberg.

It has proven difficult to construct quantum models of gravity, the remaining fundamental force. Semi-classical approximations are workable, and have led to predictions such as Hawking radiation.

However, the formulation of a complete theory of quantum gravity is hindered by apparent incompatibilities between general relativity, the most accurate theory of gravity currently known, and some of the fundamental assumptions of quantum theory.

The resolution of these incompatibilities is an area of active research, and theories such as string theory are among the possible candidates for a future theory of quantum gravity.

DERIVATION OF QUANTIZATION

The particle in a 1-dimensional potential energy box is the most simple example where restraints lead to the quantization of energy levels. The box is defined as zero potential energy inside a certain interval and infinite everywhere outside that interval. For the 1-dimensional case in the x direction, the time-independent Schrödinger equation can be written as:

$$-\frac{\hbar^2}{2m}\frac{d^2\psi}{dx^2} = \psi.$$

The general solutions are:

$$y - Ae^{ikx} + Be^{-ikx} \; E = \frac{k^2\hbar^2}{2m}$$

y = C sin kx + D cos kx (exponential rewrite)

The presence of the walls of the box restricts the acceptable solutions to the wavefunction. At each wall:

$$\psi = 0 \text{ at } x = 0, x = L$$

Consider $x = 0$

- sin 0 = 0, cos 0 = 1. To satisfy $\psi = 0$ $D = 0$ (cos term is removed)

Now Consider: $\psi = C \sin kx$

- at $X = L$, $\psi = C \sin kL$
- If C = 0 then for all x and would conflict with Born interpretation
- Therefore sin kL must be satisfied by

In this situation, n must be an integer showing the quantization of the energy levels.

Applications

Quantum mechanics has had enormous success in explaining many of the features of our world. The individual behaviour of the subatomic particles that make up all forms of matter—electrons, protons, neutrons, photons and others—can often only be satisfactorily described using quantum mechanics. Quantum mechanics has strongly influenced string theory, a candidate for a theory of everything and the multiverse hypothesis. It is also related to statistical mechanics.

Quantum mechanics is important for understanding how individual atoms combine covalently to form chemicals or molecules. The application of quantum mechanics to chemistry is known as quantum chemistry. Quantum mechanics can in principle mathematically describe most of chemistry. Quantum mechanics can provide quantitative insight into ionic and covalent bonding processes by explicitly showing which molecules are energetically favorable to which others, and by approximately how much.

Most of the calculations performed in computational chemistry rely on quantum mechanics.Much of modern technology operates at a scale where quantum effects are significant. Examples include the laser, the transistor, the electron microscope, and magnetic resonance imaging. The study of semiconductors led to the invention of the diode and the transistor, which are indispensable for modern electronics.Researchers are currently seeking robust methods of directly manipulating quantum states.

Efforts are being made to develop quantum cryptography, which will allow guaranteed secure transmission of information. A more distant goal is the development of quantum computers, which are expected to perform certain computational tasks exponentially faster than classical computers.

Another active research topic is quantum teleportation, which deals with techniques to transmit quantum states over arbitrary distances.In many devices, even the simple light switch, quantum tunneling is vital, as otherwise the electrons in the electric current could not penetrate the potential barrier made up, in the case of the light switch, of a layer of oxide. Flash memory chips found in USB drives also use quantum tunneling to erase their memory cells.

PHILOSOPHICAL CONSEQUENCES

Since its inception, the many counter-intuitive results of quantum mechanics have provoked strong philosophical debate and many interpretations. Even fundamental issues such as Max Born's basic rules concerning probability amplitudes and probability distributions took decades to be appreciated.The Copenhagen interpretation, due largely to the Danish theoretical physicist Niels Bohr, is the interpretation of quantum mechanics most widely accepted amongst physicists. According to it, the probabilistic nature of quantum mechanics predictions cannot be explained in terms of some other deterministic theory, and does not simply reflect our limited knowledge. Quantum mechanics provides probabilistic results because the physical universe is itself probabilistic rather than deterministic.

Albert Einstein, himself one of the founders of quantum theory, disliked this loss of determinism in measurement. He held that there should be a local hidden variable theory underlying quantum mechanics and consequently the present theory was incomplete.

He produced a series of objections to the theory, the most famous of which has become known as the EPR paradox. John Bell showed that the EPR paradox led to experimentally testable differences between quantum mechanics and local realistic theories. Experiments have been performed confirming the accuracy of quantum mechanics, thus demonstrating that the physical world cannot be described by local realistic theories.The writer C. S. Lewis viewed quantum mechanics as incomplete.

Lewis, a professor of English, was of the opinion that the Heisenberg uncertainty principle was more of an epistemic limitation than an indication of ontological indeterminacy, and in this respect believed similarly to many advocates of hidden variables theories.

The *Bohr-Einstein debates* provide a vibrant critique of the Copenhagen Interpretation from an epistemological point of view.The Everett many-worlds interpretation, formulated in 1956, holds that all the possibilities described by quantum theory simultaneously occur in a "multiverse" composed of mostly independent parallel universes.

This is not accomplished by introducing some new axiom to quantum mechanics, but on the contrary by *removing* the axiom of the collapse of the wave packet: All the possible consistent states of the measured system and the measuring apparatus are present in a *real* physical quantum superposition. While the multiverse is deterministic, we perceive non-deterministic behaviour governed by probabilities, because we can observe only the universe, i.e. the consistent state contribution to the mentioned superposition, we inhabit.

Everett's interpretation is perfectly consistent with John Bell's experiments and makes them intuitively understandable. However, according to the theory of quantum decoherence, the parallel universes will never be accessible to us. This inaccessibility can be understood as follows: once a measurement is done, the measured system becomes entangled with both the physicist who measured it and a huge number of other particles, some of which are photons flying away towards the other end of the universe; in order to prove that the wave function did not collapse one would have to bring all these particles back and measure them again, together with the system that was measured originally. This is completely impractical, but even if one can theoretically do this, it would destroy any evidence that the original measurement took place.

QUANTUM NUMBER

Quantum numbers describe values of conserved numbers in the dynamics of the quantum system. They often describe specifically the energies of electrons in atoms, but other possibilities include angular momentum, spin etc. Since any quantum system can have one or more quantum numbers, it is a futile job to list all possible quantum numbers.

The question of *how many quantum numbers are needed to describe any given system* has no universal answer, although for each system one must find the answer for a full analysis of the system. The dynamics of any quantum system are described by a quantum Hamiltonian, H.

There is one quantum number of the system corresponding to the energy, i.e., the eigenvalue of the

Hamiltonian. There is also one quantum number for each operator O that commutes with the Hamiltonian (i.e. satisfies the relation OH = HO). These are all the quantum numbers that the system can have. Note that the operators O defining the quantum numbers should be independent of each other. Often there is more than one way to choose a set of independent operators. Consequently, in different situations different sets of quantum numbers may be used for the description of the same system.

SINGLE ELECTRON IN AN ATOM

The most widely studied set of quantum numbers is that for a single electron in an atom: because it is not only useful in chemistry, being the basic notion behind the periodic table, valence and a host of other properties, but also because it is a solvable and realistic problem, and, as such, finds widespread. In non-relativistic quantum mechanics the Hamiltonian of this system consists of the kinetic energy of the electron and the potential energy due to the Coulomb force between the nucleus and the electron.

The kinetic energy can be separated into a piece which is due to angular momentum, J, of the electron around the nucleus, and the remainder. Since the potential is spherically symmetric, the full Hamiltonian commutes with J^2. J^2 itself commutes with any one of the components of the angular momentum vector, conventionally taken to be J_z. These are the only mutually commuting operators in this problem; hence, there are three quantum numbers.

These are conventionally known as

- The principal quantum number (n = 1, 2, 3, 4...) denotes the eigenvalue of H with the J^2 part removed. This number therefore has a dependence only on the distance between the electron and the nucleus (ie, the radial coordinate, r). The average distance increases with n, and hence quantum states with different principal quantum numbers are said to belong to different shells.
- The azimuthal quantum number (l = 0, 1... n"1) (also

known as the angular quantum number or orbital quantum number) gives the orbital angular momentum through the relation $L^2 = \hbar^2 l(l + 1)$. In chemistry, this quantum number is very important, since it specifies the shape of an atomic orbital and strongly influences chemical bonds and bond angles. In some contexts, l=0 is called an s orbital, l=1, a p orbital, l=2, a d orbital and l=3, an f orbital.

- The magnetic quantum number (m_l = "l, "l+1... 0... l"1, l) is the eigenvalue, $L_z = m_l\hbar$. This is the projection of the orbital angular momentum along a specified axis.

Results from spectroscopy indicated that up to two electrons can occupy a single orbital. However two electrons can never have the same exact quantum state nor the same set of quantum numbers according to Hund's Rules, which addresses the Pauli exclusion principle.

A fourth quantum number with two possible values was added as an *ad hoc* assumption to resolve the conflict; this supposition could later be explained in detail by relativistic quantum mechanics and from the results of the renown Stern-Gerlach experiment.

- The spin projection quantum number (m_s = "1/2 or +1/2), the intrinsic angular momentum of the electron. This is the projection of the spin s=1/2 along the specified axis.

Example: The quantum numbers used to refer to the outermost valence electron of the Fluorine (F) atom, which is located in the 2p atomic orbital, are; $n = 2$, $l = 1$, $m_l = 1$, or 0, or "1, m_s = "1/2 or 1/2.

Note that molecular orbitals require totally different quantum numbers, because the Hamiltonian and its symmetries are quite different.

Quantum Numbers with Spin-orbit Interaction

When one takes the spin-orbit interaction into consideration, l, m and s no longer commute with the Hamiltonian, and their value therefore changes over time. Thus another set of quantum numbers should be used. This set includes

- The total angular momentum quantum number (j = 1/2,3/2... n"1/2) gives the total angular momentum through the relation $J^2 = \hbar^2 j(j+1)$.
- The projection of the total angular momentum along a specified axis (m_j = -j,-j+1... j), which is analogous to m, and satisfies $m_j = m_l + m_s$.
- Parity. This is the eigenvalue under reflection, and is positive (i.e. +1) for states which came from even l and negative (i.e. -1) for states which came from odd l. The former is also known as even parity and the latter as odd parity

For example, consider the following eight states, defined by their quantum numbers:

- (1) $l = 1$, $m_l = 1$, $m_s = +1/2$
- (2) $l = 1$, $m_l = 1$, $m_s = -1/2$
- (3) $l = 1$, $m_l = 0$, $m_s = +1/2$
- (4) $l = 1$, $m_l = 0$, $m_s = -1/2$
- (5) $l = 1$, $m_l = -1$, $m_s = +1/2$
- (6) $l = 1$, $m_l = -1$, $m_s = -1/2$
- (7) $l = 0$, $m_l = 0$, $m_s = +1/2$
- (8) $l = 0$, $m_l = 0$, $m_s = -1/2$

The quantum states in the system can be described as linear combination of these eight states.

However, in the presence of spin-orbit interaction, if one wants to describe the same system by eight states which are eigenvectors of the Hamiltonian, we should consider the following eight states:

- $j = 3/2$, $m_j = 3/2$, odd parity (coming from state (1))
- $j = 3/2$, $m_j = 1/2$, odd parity (coming from states (2) and (3))
- $j = 3/2$, $m_j = -1/2$, odd parity (coming from states (4) and (5))
- $j = 3/2$, $m_j = -3/2$, odd parity (coming from state (6))
- $j = 1/2$, $m_j = 1/2$, odd parity (coming from state (2) and (3))
- $j = 1/2$, $m_j = -1/2$, odd parity (coming from states (4r) and (5))

- $j = 1/2$, $m_j = 1/2$, even parity (coming from state (7))
- $j = 1/2$, $m_j = -1/2$, even parity (coming from state (8))

ELEMENTARY PARTICLES

For a more complete description of the quantum states of elementary particles. Elementary particles contain many quantum numbers which are usually said to be intrinsic to them. However, it should be understood that the elementary particles are quantum states of the standard model of particle physics, and hence the quantum numbers of these particles bear the same relation to the Hamiltonian of this model as the quantum numbers of the Bohr atom does to its Hamiltonian.

In other words, each quantum number denotes a symmetry of the problem. It is more useful in field theory to distinguish between spacetime and internal symmetries.Typical quantum numbers related to spacetime symmetries are spin the parity, C-parity and T-parity. Typical internal symmetries are lepton number and baryon number or the electric charge. It is worth mentioning here a minor but often confusing point. Most conserved quantum numbers are additive. Thus, in an elementary particle reaction, the sum of the quantum numbers should be the same before and after the reaction. However, some, usually called a *parity*, are multiplicative; ie, their product is conserved.

All multiplicative quantum numbers belong to a symmetry in which applying the symmetry transformation twice is equivalent to doing nothing. These are all examples of an abstract group called Z_2.

Chapter 6

Quantum Nature of Light and Matter

We understand classical mechanical motion of particles governed by Newton's law. We understand the occurance of guided traveling modes and of resonator modes. There are characteristic dispersion relations or resonance frequencies associated with that. Some experimental findings at the turn of the 19th century that ultimately lead to the discover of quantum mechanics, which is that matter has in addition to its particle like properties wave properties and electromagnetic waves have in addition to its wave properties particle like properties.

As turns out the final theory, which will be developed much more than just that because the quantum mechanical wave function has a different physical interpretation than a electromagnetic wave only the mathematical concepts used is in many cases very similar. However, this is a tremendous help and guideance in doing and finally understanding quantum mechanics.

BLACK BODY RADIATION

In 1900 the physicist Max Planck found the law that governs the emission of electromagnetic radiation from a black body in thermal equilibrium. More specifically Planck's law gives the energy stored in the electromagntic field in a unit volume and unit frequency range

$[f, f + df]$ with $df = 1Hz$, when the electromagnetic field is in thermal equilibrium with its surrounding that is at temperature T. A black body is simply defined as an object that absorbs all light. The best implementation of a black body is the Ulbricht sphere.

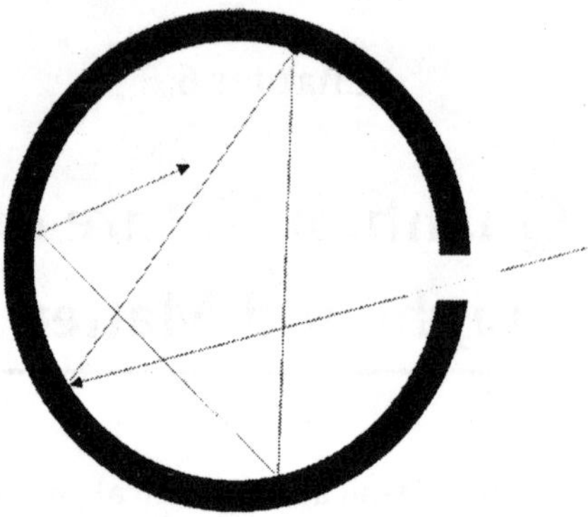

Fig. The Ulbricht Sphere

The energy density w(f) of electromagnetic radiation in a black body at temperature T.

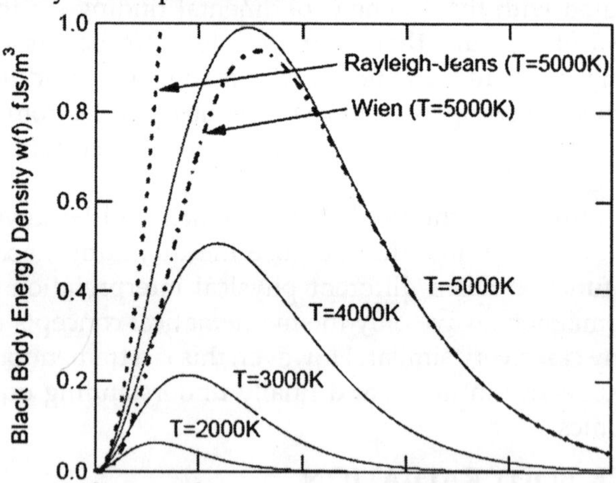

Fig. Spectral Energy Density of the black body Radiation According to Planck's Law.

Around the turn of the 19th century w(f) was measured with high precision and one was able to distinguish between various approximations that were presented by other researchers earlier, like the Rayleigh-Jeans law and Wien's law, which turned out to be asymptotic approximations to Planck's Law for low and high frequencies.

Planck had to introduce the hypothesis that harmonic oscillators with frequency f can not exchange arbitrary amounts of energy but rather only in discrete portions, so called quanta. Planck modelled atoms as classical oscillators with frequency f . Therefore, the energy of an oscillator must be quantized in energy levels corresponding to these energy quanta, which he found to be equal to *hf*, where *h* is Planck's constant

$$h = 6.62620 \pm 5.10^{-34}\ Js.$$

As a model for a black body we use now a cavity with perfectly reflecting walls, somewhat different from the Ulbricht sphere. In order to tap of a small but negligible amount of radiation from the inside, a small opening is in the wall. We can make this opening so small that it does essentially not change the internal radiation field. Then the radiation in the cavity is the sum over all possible resonator modes in the cavity. If the cavity is at temperature T all the modes are thermally excited by emission and absorption of energy quanta from the atoms of the wall.

For the derivation of Planck's law we consider a cavity with perfectly conducting walls.

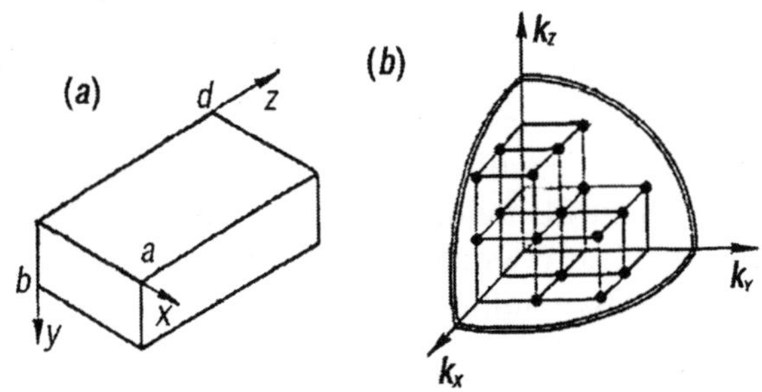

Fig. (a) Cavity Resonator with Metallic Walls.
(b) Resonator Modes Characterized by a Certain k-vector.

If we extend the analysis of the plan parallel mirror waveguide to find the TE and TM modes of a three-dimensional metalic resonator, the resonator modes are

TEmnp- and THmnp- modes characterized by its wave vector components in x-,y-, and z-direction. The resonances are standing waves in three dimensions

$$k_x = \frac{m\pi}{L_x}, k_y = \frac{n\pi}{L_y}, k_z = \frac{p\pi}{L_z}, \text{for } m, n, p = 0, 1, 2, \cdots$$

An expression for the number of modes in a frequency interval $[f, f + df]$ can be found by recognizing that this is identical to the number of points that are in the firstoctantofaspherical shell with thickness dk at $k = 2\pi f/c$ Thevolumeoccupied by onemodeinthe spaceofwave numbers k is $\Delta V = \frac{\pi}{L_x} \cdot \frac{\pi}{L_y} \cdot \frac{\pi}{L_z} = \frac{\pi^3}{V}$ with the volume $V = L_x L_y L_z$. Then the number of modes dN in the frequency interval $[f, f + df]$ in volume V are

$$dN = 2 \cdot \frac{4\pi k^2 dk}{8\frac{\pi^3}{V}} = V\frac{k^2 dk}{\pi^2},$$

where the factor of 2in front accounts for the two polarizations or TE and TH-modes of the resonator and the 8 in the denominator accounts for the fact that only one eighth of the sphere, an octand, is occupied by the positive wave vectors.

With $k = 2\pi f / c \; and \; dk = 2\pi df / c$ we obtain for the number of modes finally

$$dN = V\frac{8\pi}{c^3} f^2 df$$

Note, that the same density of states is obtained using periodic boundary conditions in all three dimensions, i.e. then we can represent all fields in terms of a three dimensional Fourier series. The possible wave vectors would range from negative to positive values

$$k_x = \frac{2m\pi}{L_x}, k_y = \frac{2n\pi}{L_y}, k_z = \frac{2p\pi}{L_z} \quad \text{for } m, n, p = 0, \pm 1, \pm 2...$$

However, these wavevectors fill the whole sphere and not just one 8-th, which compensates for the 8-times larger volume occupied by one mode. If we imply periodic boundary conditions, we have forward and backward running waves that are independent from each other. If we use the boundary conditions of the resonator, the forward and backward running waves are connected and not independent and form standing waves.

One should not be disturbed by this fact as all volume properties, such as the energy density, only depends on the density of states, and not on surface effects,aslong asthe volume is reasonably large.

RAYLEIGH-JEANS-LAW

The excitation amplitude of each mode obeys the equation of motion of a harmonic oscillator. Therefore, classically one expects that each of mode is in thermal equilibrium excute with a thermal energy kT accordingtothe equipartition theorem, where k is Boltzmann's constant with $k = 1.38062 \pm 6{\cdot}10^{-23}$J/K

If that is the case the spectral energy density is given by the Rayleigh-Jeans-Law

$$w(f) = \frac{1}{V}\frac{dN}{df}kT = \frac{8\pi}{c^3}f^2kT.$$

this law describes very well the black body radiation for frequencies $hf \ll kT$ but there is an arbitrary large deviation for high frequencies.

This formula can not be correct, because it predicts infinite energy density for the high frequency modes resulting in an "ultraviolet catastrophy", i.e. the electromagnetic field containsaninfinite amount of energy at thermal equilibrium.

WIEN'S LAW

The high frequency or short wavelength region of the black body radiation was first empirically described by Wien's Law

$$w(f) = \frac{8\pi hf^3}{c^3}e^{-hf/kT}.$$

Wien's law is surprisingly close to Planck's law, however it slightly fails to correctly predicts the asympthotic behaviour at low frequencies or long wavelengths.

PLANCK'S LAW

In the winter of 1900, Max Planck found the correct law for the black body radiation by assuming that each oscillator can only exchange energy in discrete portions or quanta. We rederive it by assuming that each mode can only have the discrete energie values.

$$E_s = s \cdot hf, \text{ for } s = 0, 1, 2, \ldots$$

Thus s is the number of energy quanta stored in the oscillator. If the oscillator is a mode of the electromagnetic field we call s the number of photons. For the probability ps, that the oscillator has the energy Es we assume a Boltzmann distribution

$$p_s = \frac{1}{Z}\exp\left(-\frac{E_s}{kT}\right) = \frac{1}{Z}\exp\left(-\frac{hf}{kT}s\right)$$

where Z is a normalization factor such that the total propability of the oscillator to have any of the allowed energy values is

$$\sum_{s=0}^{\infty} p_s = 1.$$

Note, due to the fact th∠e oscillator energy is proportional to the number of photons, the statistics are exponential statistics. We obtain for the normalization factor

$$Z = \sum_{s=0}^{\infty} \exp\left(-\frac{hf}{kT}s\right) = \frac{1}{1-\exp\left(-\frac{hf}{kT}\right)}$$

which is also called the partition function. The photon statistics are then given by

$$p_s = \exp\left(-\frac{hf}{kT}\right) = \left[1-\exp\left(-\frac{hf}{kT}\right)\right]^{-1}$$

or with $\beta = \frac{hf}{kT}$

$$p_s = \frac{1}{Z(\beta)}e^{-\beta s}, \text{ with } Z(\beta) = \sum_{s=0}^{\infty} e^{-\beta s} = \frac{1}{1-e^{-\beta}}$$

Given the statistics of the photon number, we can compute moments of the probability distribution, such as the average number of photons in the mode

$$\langle s^1 \rangle = \sum_{s=0}^{\infty} s^1 p_s$$

This first moment of the photon statistics can be computed from the partition function, using the "trick"

$$\langle s^1 \rangle = \frac{1}{Z(\beta)} \frac{\partial^1}{\partial(-\beta)^1} Z(\beta) = Z(\beta)^{e-\beta}$$

which is

$$\langle s \rangle = \frac{1}{\exp\frac{hf}{kT} - 1}.$$

With the average photon number $\langle s \rangle$ we obtain for the average energy stored in the mode $\langle E_s \rangle = \langle s \rangle hf$, and the energy density in the frequency interval $[f, f + df]$ is then given by

$$w(f) = \langle E_s \rangle \frac{dN}{df}$$

With the density of modes from Eq. we find Planck's law for the black body radiation

$$w(f) = \frac{8\pi f^2}{c^3} \frac{hf}{\exp\frac{hf}{kT} - 1}$$

which was used to make the plots. In the limits of low and high frequencies i.e. $hf << kT$ and $hf >> kT$ respectively Planck's law asympthotically approaches the Rayleigh-Jeans law and Wien's law.

THERMAL PHOTON STATISTICS

It is interesting to further investigate the intensity fluctuations of the thermal radiation emitted from a black body. If the wall opening in the Ulbricht sphere is small enough very little radiation escapes through it.

If the Ulbricht sphere is kept at constant temperature the radiation inside the Ulbricht sphere stays in thermal equilibrium and the intensity of the radiation emitted from the wall opening in a frequency interval $[f, f + df]$ is

$$I(f) = c \cdot w(f)$$

Thus the intensity fluctuations of the emitted black body radiation is directly related to the photon statistics or quantum statistics of the radiation modes at freuqency f , i.e. related to the stochastic variable s : the number of photons in a mode with frequency f . This gives us direclty experimental access to the photon statistics of an ensemble of modes or even a single mode when proper spatial and spectral filtering is applied.

Using the expectation value of the photon number 3.17, we can rewrite the photon statistics for a thermally excited mode in terms of its average photon number in the mode as

$$p_s = \frac{\langle s \rangle^s}{(\langle s \rangle + 1)^{s+1}} = \frac{1}{(\langle s \rangle + 1)} \left(\frac{\langle s \rangle}{(\langle s \rangle + 1)} \right)^s,$$

The thermal photon statistics display an exponential distribution. Before we move on, lets see how the average photon number in a given mode depends on temperature and the frequency range considered. The relationship between average number of photons in a mode with frequency f or wavelength » and temperature T.

At room temperature and micorwave frequencies large numbers of photons are present due to the thermal excitation of the mode. This is the reason that at room temperature the thermal noise overwhelms eventual quantum fluctuations. However, quantum fluctuations are importantathigh frequencies, which startfor room temperaturein the farto mid infrared range, where on average much less than one photon is thermally excited. The variance of the photon number distribution is

$$\sigma_s^2 = \langle s^2 \rangle - \langle s \rangle^2 .$$

By generalizing Eq. (3.12) to the m-th moment by replacing the exponent 1 by m

$$\langle s^m \rangle = \sum_{s=0}^{\infty} s^m p_s$$

$$= \frac{1}{Z}\frac{\partial}{\partial(-\beta)^m} Z(\beta),$$

we obtainfor the second moment

$$\langle s^m \rangle = 2Z(\beta)^2 e^{-2\beta} - Z(\beta)^2 e^{-2\beta} = 2\left\langle s^2 \right\rangle + \langle s \rangle$$

and therefore for the variance fo the photon number using Eq. is

$$\sigma_s^2 = \langle s \rangle^2 + \langle s \rangle .$$

As expected from the wide distribution of photon numbers the variance is larger than the square of the expectation value. This means that if we look at the light intensity of a single mode the intensity is subject to extremly strong fluctuations as large as the mean value. So why don't we see this rapid thermal fluctuations when we look at the black body radiation coming, for example, from the surface of the sun? Well we don't look at a single mode but rather at a whole multitude of modes. Even when we restrict us to a certain narrow frequency range and spatial direction, there is a multitude of transverse modes presence. We obtain for the average total number of photons in a group of modes and its variance

$$\langle s_{tot} \rangle = \sum_{i=1}^{N} \langle s_i \rangle ,$$

$$\sigma_{tot}^2 = \sum_{i=1}^{N} \sigma_i^2 .$$

Since these modes are independent identical systems, we have

$$\langle s_{tot} \rangle = N \times \langle s \rangle .$$

$$\sigma_{tot}^2 = N(\langle s \rangle^2 + \langle s \rangle) = \frac{1}{N}\langle s_{tot} \rangle^2 + \langle s_{tot} \rangle .$$

Due to the averaging over many modes, the photon number fluctuations in a large number of modes is reduced compared to its mean value

$$SNR = \frac{\sigma_{tot}^2}{\langle s_{tot} \rangle^2} = \frac{1}{N} + \frac{1}{\langle s_{tot} \rangle} .$$

Thus if one averages over many modes and has many photons in these modes the intensity fluctuations become small.

MODE COUNTING

It is interested to estimate the number of modes one is averaging over given a certain emitting surface and a certain measurement time.

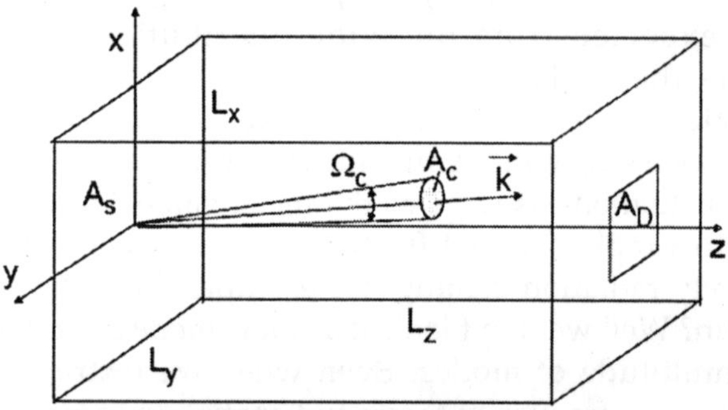

Fig. Counting of Longitudinal and Transverse modes Excited from a Radiating Surface of size as.

If the area As is emitting light, it will couple to the modes of the free field. To count the modes we put a large box (universe) over the experimental arrangement under consideration. The emitting surface is one side of the box. The light from this surface, i.e. specifying the transverse electric and magnetic fields, couples to the modes of the universe with wave vectors according to Eq.

Longitudinal Modes

The number of longitudinal modes, that propagate along the positive z-direction in the frequency interval Δf can be derived from $\Delta k = (2\pi/L_z)\ \Delta N$ and $\Delta k = (2p/c_0)\ \Delta f$

$$DN = \frac{L_z}{c_0}\Delta f,$$

or using the propagation or measurement time over which the experiment extends

$$\tau = L_z/c_0$$

we obtain for the number of longitudinal modes that are involved in the measurement that is carried out over a time intervall τ and a frequency range Δf

$$\Delta N = \tau \Delta f.$$

Transverse Modes

Thefreespace modesthatarriveatthe detector area AD will not only have wave vectors with a z- component, but also transverse components. Lets assume that the detector area is far from the emitting surface, and we consider only the paraxial plane waves. The wave vectors of these waves at a given frequency or free space wave number k0 can be approximated by

$$\vec{k}_{mn} = \left(\frac{2\pi m}{L_x}, \frac{2\pi n}{L_y}, k_0 \right) \text{ with } m, n = 0, 1, 2...$$

where m and n are transverse mode indices. Then one mode occupies the volume angle

$$\Omega_c = \frac{4\pi^2}{L_x L_y k_0^2},$$

$$= \lambda_0^2 / A_s.$$

If the modes are thermally excited, the radiation in individual modes is uncorrelated. Therefore, if there is a detector at a distance r then only the field within an area

$$A_c = r^2 \Omega_c,$$

is correlated. If the photodetector has an area A_d, then the number of transverse modes detected is

$$N_t = A_d / A_c.$$

The total number of modes detected is

$$N_{tot} = \frac{A_d}{A_c} \tau \Delta f = \frac{A_d A_s}{r^2 \lambda_0^2} \tau \Delta f.$$

Note, that there is perfect symmetry between the area of the emitting and receiving surface. The emitter and the receiver could both be black bodies. If oneofthem isatahigher temperaturethanthe other, thereisanet flow of energy from the warmer body to the colder body until equilibrium is reached.

This would not be possible without interaction over the same number of modes. Thus the formula which is completely unrelated to thermodynamics is necessary to fulfill one of the main theorems of thermodynamcis, that is that energy flows from warmer to colder bodies.

PHOTO-ELECTRIC EFFECT

Another strong indication for the quantum nature of light was the photoelectric effect by Lenard in 1903. He discovered that when ultra violet light is radiated on a photo cathode electrons are emitted

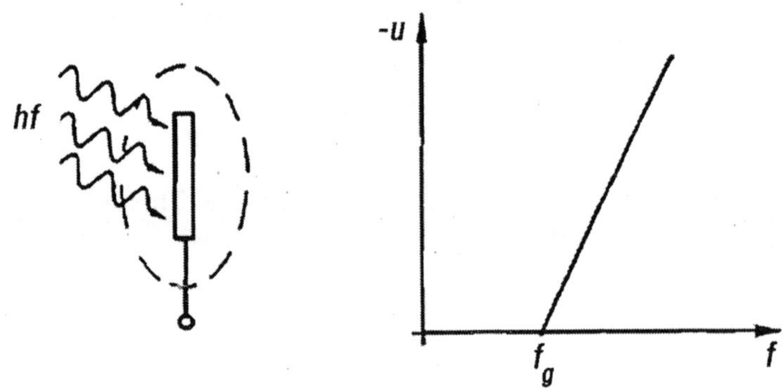

Fig. Photo-Electric Effect: (a) Schematic Setup and (b) Dependence of the Necessary Grid Voltage to Supress the Electron current as a Funtion of light Frequency.

Lenard surrounded the photo cathode by a grid, which is charged by the emitted electrons up to a voltage U, which blocks the emission of further electrons. The blocking voltage as a function of the frequency of the incoming light. Depending on the cathode material there is a cutoff frequency. For lower frequencies no electrons are emitted at all. This frequency as well as the blocking voltage does not depend on the intensity of the light.

In 1905, this effect was explained by Einstein introducing the quantum hypothesis for radiation. According to him, each electron emission is caused by a light quantum, now called photon. This photon has an energy hf and this quantum energy must be larger than the work function We of the material. The remainder of the energy mev^2/2 is transfered to the electron in form of kinetic energy. The resulting energy balance is

$$hf = We + \frac{1}{2}m_e v^2$$

The kinetic energy of the electron can be used to reach the grid surrounding the photo cathode until the charging energy due to the grid potential is equal to the kinetic energy of the electrons

$$eU = \frac{1}{2}m_e v^2 hf = We + \frac{1}{2}m_e v^2$$

or

$$-U = \frac{1}{e}(hf - We), \text{ for } hf > We.$$

This equation explains the empirically found law by Lenard explaining the cutoff frequency and the charge buildup as a function of light frequency. Einstein was firsttointroduce theideathatthe electromagnetic field contains light quanta or photons.

SPONTANEOUS AND INDUCED EMISSION

The number of photons in a radiation mode may change via emission of photon into the mode or absorption of a photon from the mode by atoms, molecules or a solid state material. Einstein introduced a phenomenological theory of these processes in order to explain how matter may get into thermal equilibrium by interaction with the modes of the radiation field. He considered the interaction of a mode with atoms modeled by two energy levels E_1 and E_2.

n_1 and n_2 are the population densities of these two levels considering a whole ensemble of these atoms. Transitions are possible in the atom between the two energy levels by emission of a photon at a frequency

$$f = \frac{E_2 - E_1}{h}$$

Absorption of a photon is only possible if there is energy present in the radiation field.

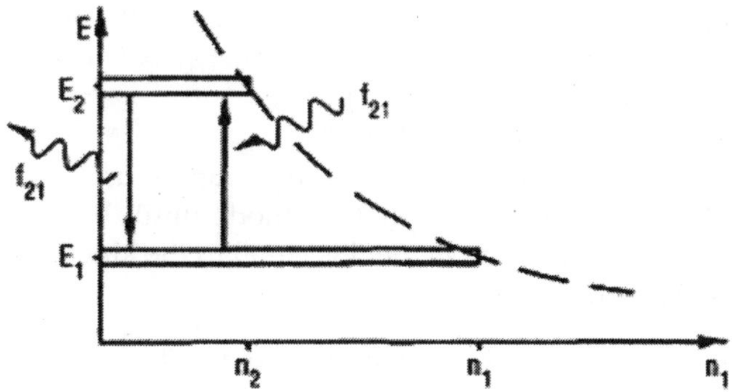

Fig. Energy Levels of a two Level Atom and Populations.

Einstein wrote for the corresponding transition rates, which should be proportional to the population densities and the photon density at the transition frequency

$$-\left.\frac{dn_1}{dt}\right|_{Abs} = \left.\frac{dn_2}{dt}\right|_{Abs} = B_{12}n_1w(f_{21}).$$

The coefficient B_{12} characterizes the absorption properties of the transition. Einstein had to allow for two different kind of processes for reasons that become clear a little later. Transitions induced by the already present photons or radiation energy as well as spontaneous transitions

$$\left.\frac{dn_1}{dt}\right|_{Em} = -\left.\frac{dn_2}{dt}\right|_{Em} = B_{12}n_2w(f_{21}) + A_{12}\, n_2.$$

The coefficient B_{21} escribes the induced and A_{21} the spontaneous emissions. The latter transitions occur even in the absence of any radiation and the corresponding coefficient determines the lifetime of the excited state

$$\tau_{sp} = A_{21}^{-1},$$

in the absence of the radiation field. The total change in the population densities is due to both absorption and emission processes

$$\frac{dn_i}{dt} = \left.\frac{dn_i}{dt}\right|_{Em} + \left.\frac{dn_i}{dt}\right|_{Abs} = \text{for } i = 1, 2.$$

Using Eqs. (3.44) and (3.45) we find

$$-\frac{dn_1}{dt}=\frac{dn_2}{dt}=(B_{12}n_1-B_{21}n_2)\,w(f_{21})-A_{21}n_2.$$

In thermal equilibrium the energy density of the radiation field must fulfill the condition

$$w\,(f_{21})=\frac{A_{21}/B_{12}}{n_1/n_2-B_{21}/B_{12}},$$

while theatomicensembleitself shouldalsobeinthermal equilibrium which again should be described by the Boltzmann statistics, i.e. the ratio between the population densities are determined by the Boltzmann factor

$$n_2/n_1=\exp\left(-\frac{E_2-E_1}{kT}\right).$$

And with it the energy density of the radiation field must be

$$w(f_{21})=\frac{A_{21}/B_{12}}{\exp\left(\frac{hf_{21}}{kT}\right)-B_{21}/B_{12}}.$$

A comparison with Planck's law, Eq.gives

$$B_{21}=B_{12},$$

and

$$A21=\frac{8\pi hf_{21}^3}{c^3}B_{12}.$$

Clearly, without the spontaneous emission process it is impossible to arrive at Planck's Law in equilibrium. The spectral energy density of the radiation fieldcan be rewrittenwiththe average photonnumberinthe modesatthe transition frequency f_{21} as

$$w(f_{21})=\frac{8\pi hf_{21}^2}{c^3}hf_{21}\langle s\rangle,\quad \text{with}\ \langle s\rangle=\langle s\rangle=\frac{1}{\exp\frac{hf_{21}}{kT}-1}.$$

Or we can write

$$w(f_{21})=\frac{A_{21}}{B_{12}}\langle s\rangle.$$

With that relationship Eq. can be rewritten as

$$\left.\frac{dn_1}{dt}\right|_{Em}=-\left.\frac{dn_2}{dt}\right|_{Em}=A_{21}n_2(\langle s\rangle+1),$$

which indicates that the number of spontaneous emissions is equvalent to induced emissions caused by the presence of a single photon per mode.

Having identified the coefficients describing the transition rates in the atom interacting with the field from equilibrium considerations, we can rewrite the rate equations also for the non equilibrium situation, because the coefficients are constants depending only on the transition considered

$$\frac{dn_1}{dt} = -\frac{dn_2}{dt} = \frac{1}{\tau_{sp}}[(n_2 - n_1)\langle s\rangle + n_2].$$

With each transition from the excited state of the atom to the ground state an emission of a photon goes along with it. From this, we obtain a change in the average photon number of the modes

$$\frac{d\langle s\rangle}{dt} = V\frac{dn_1}{dt},$$

which is

$$\frac{d\langle s\rangle}{dt} = \frac{V}{\tau_{sp}}[(n_2 - n_1)\langle s\rangle + n_2].$$

Again the first term describes the stimulated or induced processes and the second term the spontaneous processes. The stimulated emission processes are coherent with the already present radiation field that is inducing the transitions.

This is not so for the spontaneous emissions, which add noise to the already present field. For $n_1 > n_2$ the stimulated processes lead to a decrease in the photon number and the medium is absorbing. In the case of inversion, $n_2 > n_1$, the photon number increases exponentially.

According to Eq. inversion corresponds to a negative temperature, which is an indication for a non equilibrium situation that can only be maintained by additional means. It is impossible to achieve inversion by simple irradiation of the atoms with intense radiation from Eq in steady state the ratio between excited state and ground state population is

$$\frac{n_2}{n_1} = \frac{\langle s\rangle}{\langle s\rangle + 1},$$

which at most approaches equal population for very large photon number. However, such a process can be exploited in a three or four level system.

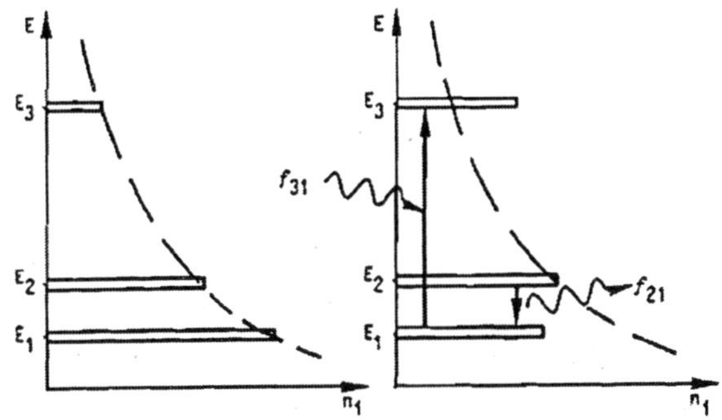

Fig. Three Level System: (a) In Thermal Equilibrium and (b) Under Optical Pumping at the Transition Frequency f_{31}.

By optical pumping population from the ground state can be transfered to the excited level with energy E_3.

If there is a fast relaxation from this level to level E_2, where level two in contrast has a long lifetime, it is conceivable that an inversion between level E_2 and E_1 can build up. If inversion is achieved radiation at the frequency f_{21} is amplified.

Chapter 7

Cosmic Rays

Galactic cosmic rays (GCRs) are the high-energy particles that flow into our solar system from far away in the Galaxy. GCRs are mostly pieces of atoms: protons, electrons, and atomic nuclei which have had all of the surrounding electrons stripped during their high-speed (almost the speed of light) passage through the Galaxy.

Cosmic rays provide one of our few direct samples of matter from outside the solar system. The magnetic fields of the Galaxy, the solar system, and the Earth have scrambled the flight paths of these particles so much that we can no longer point back to their sources in the Galaxy. If you made a map of the sky with cosmic ray intensities, it would be completely uniform. So we have to determine where cosmic rays come from by indirect means.

One of the indirect observations we can make is the "composition" of GCRs. This can tell us a lot about the GCR sources and the cosmic rays' trip through the Galaxy. The "composition" of cosmic rays describes what fraction of cosmic rays are protons, what fraction are helium nuclei, etc. All of the natural elements in the periodic table are present in cosmic rays, in roughly the same proportion as they occur in the solar system. But detailed differences provide a "fingerprint" of the cosmic ray's source.

Measuring the quantity of each different element is relatively easy, since the different charges of each nucleus give very different signatures. Harder to measure, but a better fingerprint, is the isotopic composition (nuclei of the same element but with different numbers of neutrons). To tell the isotopes apart involves, in effect, weighing each atomic nucleus that enters the cosmic ray detector.

About 90% of the cosmic ray nuclei are hydrogen (protons), about 9% are helium (alpha particles), and all of the rest of the elements make up only 1%. Even in this one percent there are very rare elements and isotopes. These require large detectors to collect enough particles to say something meaningful about the "fingerprint" of their source. The HEAO Heavy Nuclei Experiment, launched in 1979, collected only about 100 cosmic rays between element 75 and element 87 (the group of elements that includes platinum, mercury, and lead), in almost a year and a half of flight, and it was much bigger than most scientific instruments flown by NASA today. To make better measurements requires an even larger instrument, and the bigger the instrument, the greater the cost.

COSMIC RAY SOURCES

Most cosmic rays originate from extrasolar sources within our own galaxy such as rotating neutron stars, supernovae, and black holes. However, the fact that some cosmic rays have extremely high energies provides evidence that at least some must be of extra-galactic origin (e.g. radio galaxies and quasars); the local galactic magnetic field would not be able to contain particles with such a high energy.

The origin of cosmic rays with energies up to 10^{14} eV can be accounted for in terms of shock-wave acceleration in supernova shells. The origin of cosmic rays with energy greater than 10^{14} eV remained unknown until recently, when a large collaborative experiment at the Pierre Auger Observatory appears to have answered this question. In preliminary results announced in November 2007, they showed a strong correlation between their 27 most energetic events and active galactic nuclei [AGN]. These results demonstrated that there is only a small chance (less than 1/100) that the highest energy protons originated from outside the AGN.

Observations have shown that cosmic rays with an energy above 10 GeV (10×10^9 eV) approach the Earth's surface isotropically (equally from all directions); it has been hypothesized that this is not due to an even distribution of cosmic ray sources, but instead is due to galactic magnetic fields causing cosmic rays to travel in spiral paths. This limits cosmic ray's usefulness in positional astronomy as they carry no information of their direction of origin. At energies below 10 GeV there is a directional dependence, due to the interaction of the charged component of the cosmic rays with the Earth's magnetic field.

Solar Cosmic Rays

Solar cosmic rays or solar energetic particles (SEP) are cosmic rays that originate from the Sun. The average composition is similar to that of the Sun itself. There exists no clear and sharp boundary between the phase spaces of the solar wind and SEP plasma particle populations.

The name solar cosmic ray itself is a misnomer because the term cosmic implies that the rays are from the cosmos and not the solar system, but it has stuck. The misnomer arose because there is continuity in the energy spectra, i.e., the flux of particles as a function of their energy, because the low-energy solar cosmic rays fade more or less smoothly into the galactic ones as one looks at increasingly higher energies. Until the mid-1960s the energy distributions were generally averaged over long time intervals, which also obscured the difference.

Later, it was found that the solar cosmic rays vary widely in their intensity and spectrum, increasing in strength after some solar events such as solar flares. Further, an increase in the intensity of solar cosmic rays is followed by a decrease in all other cosmic rays, called the Forbush decrease after their discoverer, the physicist Scott Forbush.

These decreases are due to the solar wind with its entrained magnetic field sweeping some of the galactic cosmic rays outwards, away from the Sun and Earth. The overall or average rate of Forbush decreases tends to follow the 11-year sunspot cycle, but individual events are tied to events on the Sun, as explained above. There are further differences between cosmic rays of solar and galactic origin, mainly in that the galactic cosmic rays show an enhancement of heavy elements such as calcium, iron and gallium, as well as of cosmically rare light elements such as lithium and beryllium.

The latter result from the cosmic ray spallation (fragmentation) of heavy nuclei due to collisions in transit from the distant sources to the solar system.

Anomalous Cosmic Rays

Anomalous cosmic rays (ACRs) are cosmic rays with unexpectedly low energies. They are thought to be created near the edge of our solar system, in the heliosheath, the border region between the heliosphere and the interstellar medium. When electrically neutral atoms are able to enter the heliosheath (being unaffected by its magnetic fields) subsequently become ionized, they are thought to be accelerated into low-energy cosmic rays by the solar wind's termination shock which marks the inner edge of the heliosheath.

It is also possible that high energy galactic cosmic rays which hit the shock front of the solar wind near the heliopause might be decelerated, resulting in their transformation into lower-energy anomalous cosmic rays.

The *Voyager 1* space probe crossed the termination shock on December 16, 2004, according to papers published in the journal *Science*. Readings showed particle acceleration, but not of the kind that generates ACRs. It is unclear at this stage (September 2005) if this is typical of the termination shock (requiring a major rethink of the origin of ACRs), or a localised feature of that part of the termination shock that *Voyager 1* passed through. *Voyager 2* is expected to cross the termination shock during or after 2008, which will provide more data.

COMPOSITION OF COSMIC RAYS

Cosmic rays may broadly be divided into two categories, primary and secondary. The cosmic rays that arise in extrasolar astrophysical sources are primary cosmic rays; these primary cosmic rays can interact with interstellar matter to create secondary cosmic rays. The sun also emits low energy cosmic rays associated with solar flares. The exact composition of primary cosmic rays, outside the Earth's atmosphere, is dependent on which part of the energy spectrum is observed. However, in general, almost 90% of all the incoming cosmic rays are protons, about 9% are helium nuclei (alpha particles) and about 1% are electrons.

The remaining fraction is made up of the other heavier nuclei which are abundant end products of star's nuclear synthesis. Secondary cosmic rays consist of the other nuclei which are not abundant nuclear synthesis end products, or products of the Big Bang, primarily lithium, beryllium and boron. These light nuclei appear in cosmic rays in much greater abundance (about 1:100 particles) than in solar atmospheres, where their abundance is about 10^{-7} that of helium.

This abundance difference is a result of the way secondary cosmic rays are formed. When the heavy nuclei components of primary cosmic rays, namely the carbon and oxygen nuclei, collide with interstellar matter, they break up into lighter nuclei (in a process termed cosmic ray spallation), into lithium, beryllium and boron.

It is found that the energy spectra of Li, Be and B falls off somewhat steeper than that of carbon or oxygen, indicating that less cosmic ray spallation occurs for the higher energy nuclei presumably due to their escape from the galactic magnetic field. Spallation is also responsible for the abundances of Sc, Ti, V and Mn elements in cosmic rays, which are produced by collisions of Fe and Ni nuclei with interstellar matter.

In the past, it was believed that the cosmic ray flux has remained fairly constant over time. Recent research has, however, produced evidence for 1.5 to 2-fold millennium-timescale changes in the cosmic ray flux in the past forty thousand years.

MODULATION

The flux (flow rate) of cosmic rays incident on the Earth's upper atmosphere is modulated (varied) by two processes; the sun's solar wind and the Earth's magnetic field. Solar wind is expanding magnetized plasma generated by the sun, which has the effect of decelerating the incoming particles as well as partially excluding some of the particles with energies below about 1 GeV. The amount of solar wind is not constant due to changes in solar activity over its regular eleven-year cycle. Hence the level of modulation varies in autocorrelation with solar activity. Also the Earth's magnetic field deflects some of the cosmic rays, which is confirmed by the fact that the intensity of cosmic radiation is dependent on latitude, longitude and azimuth.

The cosmic flux varies from eastern and western directions due to the polarity of the Earth's geomagnetic field and the positive charge dominance in primary cosmic rays; this is termed the east-west effect. The cosmic ray intensity at the equator is lower than at the poles as the geomagnetic cutoff value is greatest at the equator. This can be understood by the fact that charged particle tend to move in the direction of field lines and not across them. This is the reason the Aurorae occur at the poles, since the field lines curve down towards the Earth's surface there. Finally, the longitude dependence arises from the fact that the geomagnetic dipole axis is not parallel to the Earth's rotation axis.

This modulation which describes the change in the interstellar intensities of cosmic rays as they propagate in the heliosphere is highly energy and spatial dependent, and it is described by the Parker's Transport Equation in the heliosphere. At large radial distances, far from the Sun ~ 94 AU, there exists the region where the solar wind undergoes a transition from supersonic to subsonic speeds called the solar wind termination shock.

The region between the termination shock and the heliospause (the boundary marking the end of the heliosphere) is called the heliosheath. This region acts as a barrier to cosmic rays and it decreases their intensities at lower energies by about 90% indicating that it is not only the Earth's magnetic field that protect us from cosmic ray bombardment.

From modelling point of view, there is a challenge in determining the Local Interstellar spectra (LIS) due to large adiabatic energy changes these particles experience owing to the diverging solar wind in the heliosphere.

However, significant progress has been made in the field of cosmic ray studies with the development of an improved state-of-the-art 2D numerical model that includes the simulation of the solar wind termination shock, drifts and the heliosheath coupled with fresh descriptions of the diffusion tensor. But challenges also exist because the structure of the solar wind and the turbulent magnetic field in the heliosheath is not well understood indicating the heliosheath as the region unknown beyond.

With lack of knowledge of the diffusion coefficient perpendicular to the magnetic field our knowledge of the heliosphere and from the modelling point of view is far from complete. There exist promising theories like ab initio approaches, but the drawback is that such theories produce poor compatibility with observations (Minnie, 2006) indicating their failure in describing the mechanisms influencing the cosmic rays in the heliosphere.

DETECTION

The nuclei that make up cosmic rays are able to travel from their distant sources to the Earth because of the low density of matter in space. Nuclei interact strongly with other matter, so when the cosmic rays approach Earth they begin to collide with the nuclei of atmospheric gases.

These collisions, in a process known as a shower, result in the production of many pions and kaons, unstable mesons which quickly decay into muons.

Because muons do not interact strongly with the atmosphere and because of the relativistic effect of time dilation many of these muons are able to reach the surface of the Earth. Muons are ionizing radiation, and may easily be detected by many types of particle detectors such as bubble chambers or scintillation detectors.

If several muons are observed by separated detectors at the same instant it is clear that they must have been produced in the same shower event.

Detection by Particle Track-etch Technique

Cosmic rays can also be detected directly when they pass through particle detectors flown aboard satellites or in high altitude balloons. In a pioneering technique developed by P. Buford Price et al., sheets of clear plastic such as 1/4 mil Lexan polycarbonate can be stacked together and exposed directly to cosmic rays in space or high altitude. When returned to the laboratory, the plastic sheets are "etched" [literally, slowly dissolved] in warm caustic sodium hydroxide solution, which slowly removes the surface material at a slow, known rate. Wherever a bare cosmic ray nucleus passes through the detector, the nuclear charge causes chemical bond breaking in the plastic.

The slower the particle, the more extensive is the bond-breaking along the path; and the higher the charge [the higher the Z], the more extensive is the bond-breaking along the path. The caustic sodium hydroxide dissolves at a faster rate along the path of the damage, but thereafter dissolves at the slower base-rate along the surface of the minute hole that was drilled. The net result is a conical shaped pit in the plastic; typically with two pits per sheet [one originating from each side of the plastic].

The etch pits can be measured under a high power microscope [typically 1600X oil-immersion], and the etch rate plotted as a function of the depth in the stack of plastic. At the top of the stack, the ionization damage is less due to the higher speed. As the speed decreases due to deceleration in the stack, the ionization damage increases along the path. This generates a unique curve for each atomic nucleus of Z from 1 to 92, allowing identification of both the charge and energy [speed] of the particle that traverses the stack. This technique has been used with great success for detecting not only cosmic rays, but fission product nuclei for neutron detectors.

Interaction with the Earth's Atmosphere

When cosmic ray particles enter the Earth's atmosphere they collide with molecules, mainly oxygen and nitrogen, to produce a cascade of lighter particles, a so-called air shower.

The general idea which shows a cosmic ray shower produced by a high energy proton of cosmic ray origin striking an atmospheric molecule.

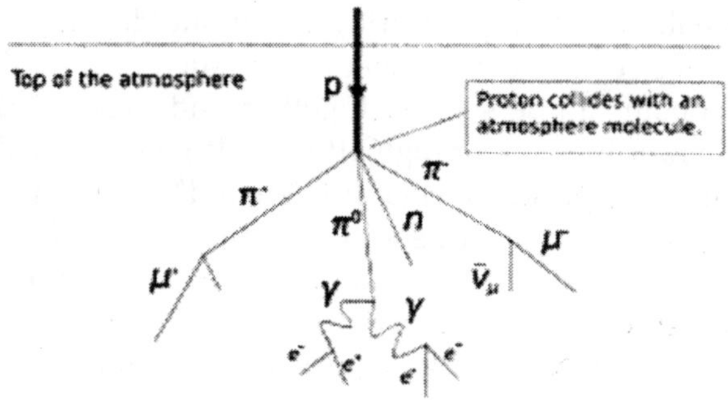

This image is a simplified picture of an air shower: in reality, the number of particles created in an air shower event can reach in the billions, depending on the energy and chemical environment (i.e. atmospheric) of the primary particle. All of the produced particles stay within about one degree of the primary particle's path. Typical particles produced in such collisions are charged mesons (e.g. positive and negative pions and kaons); one common collision is:

$$p + O^{16} \rightarrow n + \pi$$

Cosmic rays are also responsible for the continuous production of a number of unstable isotopes in the Earth's atmosphere, such as carbon-14, via the reaction:

$$n + N^{14} \rightarrow p + C^{14}$$

Cosmic rays kept the level of carbon-14 in the atmosphere roughly constant (70 tons) for at least the past 100,000 years, until the beginning of aboveground nuclear weapons testing in the early 1950s. This is an important fact used in radiocarbon dating which is used in archaeology.

UNUSUAL COSMIC RAYS

In 1975, a team of researchers headed by P. Buford Price at U.C. Berkeley announced the discovery of a cosmic ray track in a particle detector slung under a high-altitude balloon that was significantly different from all others ever measured. Using the particle track-etch method pioneered by Price, et al., they discovered the track of a particle that had passed through 32 sheets of 1/4 mil Lexan plastic without any measurable change in ionization. Yet, the Cerenkov detector admitted only of particles less than 2/3 c [the speed of light in the clear plastic]. The charge was measured as being 137, the same as predicted by Paul Dirac who first predicted the theoretical existence of magnetic monopoles. The particle track preliminarily identified as having been caused by a magnetic monopole had been spotted by technical assistant Walter L. Wagner.

A possible alternative explanation was offered by Alvarez. In his paper it was demonstrated that the path of the cosmic ray event that was claimed to be due to a magnetic monopole could be reproduced by a path followed by a platinum nucleus fragmenting to osmium and then to tantalum.There are a number of cosmic ray research initiatives. These include, but are not limited to:

- Chicos
- Pamela
- Alpha Magnetic Spectrometer
- High Resolution Fly's Eye Cosmic Ray Detector
- Mariachi
- Pierre Auger Observatory
- Spaceship Earth
- Telescope Array Project

HISTORY OF COSMIC RAYS

After the discovery of radioactivity by Henri Becquerel in 1896, it was generally believed that atmospheric electricity (ionization of the air) was caused only by radiation from radioactive elements in the ground or the radioactive gases (isotopes of radon) they produce. Measurements of ionization rates at increasing heights above the ground during the decade from 1900 to 1910 showed a decrease that could be explained as due to absorption of the ionizing radiation by the intervening air. In 1912 Domenico Pacini observed simultaneous variations of the rate of ionization over a lake, and over the sea. Pacini concluded that a certain part of the ionization must be due to sources other than the radioactivity of the Earth or the air.

Then, in 1912, Victor Hess carried three Wulf electrometers (a device to measure the rate of ion production inside a hermetically sealed container) to an altitude of 5300 meters in a free balloon flight. He found the ionization rate increased approximately fourfold over the rate at ground level. He concluded "The results of my observation are best explained by the assumption that a radiation of very great penetrating power enters our atmosphere from above." In 1913-14, Werner Kolhörster confirmed Victor Hess' results by measuring the increased ionization rate at an altitude of 9 km. Hess received the Nobel Prize in Physics in 1936 for his discovery.

The term "cosmic rays" was coined by Robert Millikan who proved they were extraterrestrial in origin, and not produced by atmospheric electricity as Hess had thought. Millikan believed that cosmic rays were high-energy photons with some secondary electrons produced by Compton scattering of gamma rays.

Compton himself held the (correct) belief that cosmic rays were primarily charged particles. During the decade from 1927 to 1937, a wide variety of experimental investigations demonstrated that the primary cosmic rays are mostly positively charged particles, and the secondary radiation observed at ground level is composed primarily of a "soft component" of electrons and photons and a "hard component" of penetrating particles, muons.

The muon was initially believed to be the unstable particle predicted by Hideki Yukawa in 1935 in his theory of the nuclear force. Experiments proved that the muon decays with a mean life of 2.2 microseconds into an electron and two neutrinos, but that it does not interact strongly with nuclei, so it could not be the Yukawa particle. The mystery was solved by the discovery in 1947 of the pion, which is produced directly in high-energy nuclear interactions. It decays into a muon and one neutrino with a mean life of 0.0026 microseconds. The pion'!muon'!electron decay sequence was observed directly in a microscopic examination of particle tracks in a special kind of photographic plate called a nuclear emulsion that had been exposed to cosmic rays at a high-altitude mountain station. In 1948, observations with nuclear emulsions carried by balloons to near the top of the atmosphere by Gottlieb and Van Allen showed that the primary cosmic particles are mostly protons with some helium nuclei (alpha particles) and a small fraction heavier nuclei.

In 1934 Bruno Rossi reported an observation of near-simultaneous discharges of two Geiger counters widely separated in a horizontal plane during a test of equipment he was using in a measurement of the so-called east-west effect. In his report on the experiment, Rossi wrote "...it seems that once in a while the recording equipment is struck by very extensive showers of particles, which causes coincidences between the counters, even placed at large distances from one another. Unfortunately, he did not have the time to study this phenomenon more closely."

In 1937 Pierre Auger, unaware of Rossi's earlier report, detected the same phenomenon and investigated it in some detail. He concluded that extensive particle showers are generated by high-energy primary cosmic-ray particles that interact with air nuclei high in the atmosphere, initiating a cascade of secondary interactions that ultimately yield a shower of electrons, photons, and muons that reach ground level.

Homi J. Bhabha derived an expression for the probability of scattering positrons by electrons, a process now known as Bhabha scattering. His classic paper, jointly with Warren Heitler, published in 1937 described how primary cosmic rays from space interact with the upper atmosphere to produce particles observed at the ground level. Bhabha and Heitler explained the cosmic ray shower formation by the cascade production of gamma rays and positive and negative electron pairs. In 1938 Bhabha concluded that observations of the properties of such particles would lead to the straightforward experimental verification of Albert Einstein's theory of relativity.

Measurements of the energy and arrival directions of the ultra-high-energy primary cosmic rays by the techniques of "density sampling" and "fast timing" of extensive air showers were first carried out in 1954 by members of the Rossi Cosmic Ray Group at the Massachusetts Institute of Technology.

The experiment employed eleven scintillation detectors arranged within a circle 460 meters in diameter on the grounds of the Agassiz Station of the Harvard College Observatory. From that work, and from many other experiments carried out all over the world, the energy spectrum of the primary cosmic rays is now known to extend beyond 10^{20} eV (past the GZK cutoff, beyond which very few cosmic rays should be observed). A huge air shower experiment called the Auger Project is currently operated at a site on the pampas of Argentina by an international consortium of physicists. Their aim is to explore the properties and arrival directions of the very highest energy primary cosmic rays.

The results are expected to have important implications for particle physics and cosmology. In November, 2007 preliminary results were announced showing direction of origination of the 27 highest energy events were strongly correlated with the locations of active galactic nuclei [AGN], where bare protons are believed accelerated by strong magnetic fields associated with the large black holes at the AGN centers to energies of 1E20 eV and higher.

Three varieties of neutrino are produced when the unstable particles produced in cosmic ray showers decay. Since neutrinos interact only weakly with matter most of them simply pass through the Earth and exit the other side. They very occasionally interact, however, and these atmospheric neutrinos have been detected by several deep underground experiments.

The Super-Kamiokande in Japan provided the first convincing evidence for neutrino oscillation in which one flavour of neutrino changes into another. The evidence was found in a difference in the ratio of electron neutrinos to muon neutrinos depending on the distance they have traveled through the air and earth.

EFFECTS OF COSMIC RAYS

Role in Ambient Radiation

Cosmic rays constitute a fraction of the annual radiation exposure of human beings on earth. For example, the average radiation exposure in Australia is 0.3 mSv due to cosmic rays, out of a total of 2.3 mSv.

Effect on Electronics

Cosmic rays have sufficient energy to alter the states of elements in electronic integrated circuits, causing transient errors to occur, such as corrupted data in memory, or incorrect behaviour of a CPU. This has been a problem in high-altitude electronics, such as in satellites, but as transistors become smaller it is becoming an increasing concern in ground-level equipment as well. To alleviate this problem, Intel has proposed a cosmic ray detector which could be integrated into future high-density microprocessors, allowing the processor to repeat the last command following a cosmic ray event.

Significance to Space Travel

Galactic cosmic rays are one of the most important barriers standing in the way of plans for interplanetary travel by crewed spacecraft.

Role in Lightning

Cosmic rays have been implicated in the triggering of electrical breakdown in lightning. It has been proposed that essentially all lightning is triggered through a relativistic process, "runaway breakdown", seeded by cosmic ray secondaries. Subsequent development of the lightning discharge then occurs through "conventional breakdown" mechanisms.

Role in Climate Change

Whether cosmic rays have any role in climate change is disputed. Different groups have made different arguments regarding the role of cosmic ray forcing in climate change.

Shaviv et al. have argued that *galactic cosmic ray* (GCR) climate signals on geological time scales are attributable to changing positions of the galactic spiral arms of the Milky Way, and that cosmic ray flux variability is the dominant climate driver over these time periods. They also argue that GCR flux variability plays an important role in climate variability over shorter time scales, though the relative contribution of anthropogenic factors in relation to GCR flux presently is a matter of continued debate. Because of uncertainty about which GCR energies are the most important drivers of cloud cover variation (if any), and because of the paucity of historical data on cosmic ray flux at various ranges of energies, controversies remain.

Henrik Svensmark et al. have argued that solar variations modulate the cosmic ray signal seen at the earth and that this would affect cloud formation and hence climate. Cosmic rays have been experimentally determined to be able to produce ultra-small aerosol particles, orders of magnitude smaller than cloud condensation nuclei (CCN). Whether this mechanism is relevant to the real atmosphere is unknown; in particular, the steps from this to modulation of cloud formation and thence climate have not been established.

The analogy is with the Wilson cloud chamber, however acting on a global scale, where earth's atmosphere acts as the cloud chamber and the cosmic rays catalyze the production of CCN. But unlike a cloud chamber, where the air is carefully purified, the real atmosphere always has many CCN naturally. Various proposals have been made for the mechanism by which cosmic rays might affect clouds, including ion mediated nucleation, and indirect effects on current flow density in the global electric circuit . Claims have been made of identification of GCR climate signals in atmospheric parameters such as high latitude precipitation and Svensmark's annual cloud cover variations, which were said to be correlated to GCR variation.

COSMIC RAYS AND FICTION

Because of the metaphysical connotations of the word "cosmic", the very name of these particles enables their misinterpretation by the public, giving them an aura of mysterious powers.

Were they merely referred to as "high-speed protons and atomic nuclei" this might not be so.In fiction, cosmic rays have been used as a catchall, mostly in comics (notably the *Marvel Comics* group the Fantastic Four), as a source for mutation and therefore the powers gained by being bombarded with them. Most galactic cosmic rays are probably accelerated in the blast waves of supernova remnants.

This doesn't mean that the supernova explosion itself gets the particles up to these speeds. The remnants of the explosions, expanding clouds of gas and magnetic field, can last for thousands of years, and this is where cosmic rays are accelerated. Bouncing back and forth in the magnetic field of the remnant randomly lets some of the particles gain energy, and become cosmic rays. Eventually they build up enough speed that the remnant can no longer contain them, and they escape into the Galaxy.

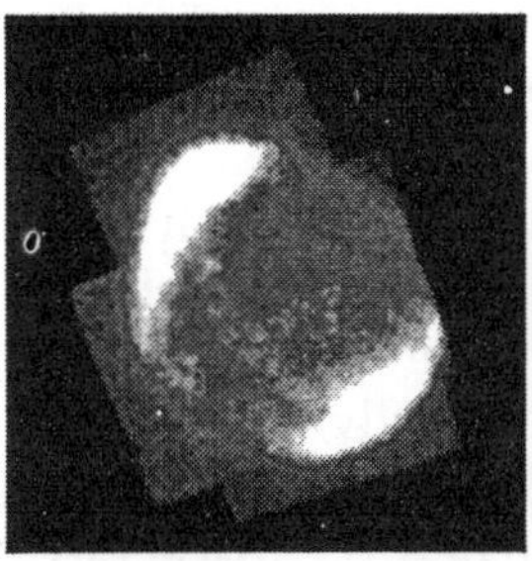

Supernova 1006

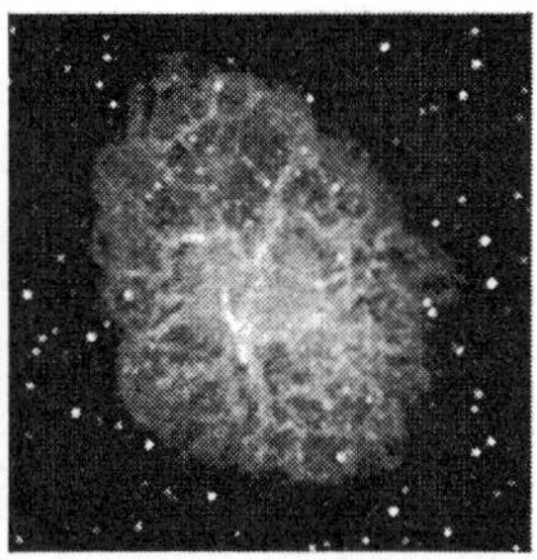

Crab Nebula

Because the cosmic rays eventually escape the supernova remnant, they can only be accelerated up to a certain maximum energy, which depends upon the size of the acceleration region and the magnetic field strength.

However, cosmic rays have been observed at much higher energies than supernova remnants can generate, and where these ultra-high-energies come from is a big question. Perhaps they come from outside the Galaxy, from active galactic nuclei, quasars or gamma ray bursts. Or perhaps they're the signature of some exotic new physics: superstrings, exotic dark matter, strongly-interacting neutrinos, or topological defects in the very structure of the universe. Questions like these tie cosmic-ray astrophysics to basic particle physics and the fundamental nature of the universe.

Cosmic rays were discovered in 1912 by Victor Hess, when he found that an electroscope discharged more rapidly as he ascended in a balloon. He attributed this to a source of radiation entering the atmosphere from above, and in 1936 was awarded the Nobel prize for his discovery. For some time it was believed that the radiation was electromagnetic in nature (hence the name cosmic "rays"). However, during the 1930's it was found that cosmic rays must be electrically charged because they are affected by the Earth's magnetic field.

From the 1930s to the 1950s, before man-made particle accelerators reached very high energies, cosmic rays served as a source of particles for high energy physics investigations, and led to the discovery of subatomic particles that included the positron and muon.

Although these applications continue, since the dawn of the space age the main focus of cosmic ray research has been directed towards astrophysical investigations of where cosmic rays originate, how they get accelerated to such high velocities, what role they play in the dynamics of the Galaxy, and what their composition tells us about matter from outside the solar system. To measure cosmic rays directly, before they have been slowed down and broken up by the atmosphere, research is carried out by instruments carried on spacecraft and high altitude balloons, using particle detectors similar to those used in nuclear and high energy physics experiments.

Cosmic Ray Energies and Acceleration

The energy of cosmic rays is usually measured in units of MeV, for mega-electron volts, or GeV, for giga-electron volts. (One electron volt is the energy gained when an electron is accelerated through a potential difference of 1 volt). Most galactic cosmic rays have energies between 100 MeV (corresponding to a velocity for protons of 43% of the speed of light) and 10 GeV (corresponding to 99.6% of the speed of light). The number of cosmic rays with energies beyond 1 GeV decreases by about a factor of 50 for every factor of 10 increase in energy. Over a wide energy range the number of particles per m^2 per steradian per second with energy greater than E (measured in GeV) is given approximately by $N(>E) = k(E + 1)^{-a}$, where $k \sim 5000$ per m^2 per steradian per second and $a \sim 1.6$. The highest energy cosmic rays measured to date have had more than 10^{20} eV, equivalent to the kinetic energy of a baseball traveling at approximately 100 mph!

It is believed that most galactic cosmic rays derive their energy from supernova explosions, which occur approximately once every 50 years in our Galaxy. To maintain the observed intensity of cosmic rays over millions of years requires that a few percent of the more than 10^{51} ergs released in a typical supernova explosion be converted to cosmic rays. There is considerable evidence that cosmic rays are accelerated as the shock waves from these explosions travel through the surrounding interstellar gas. The energy contributed to the Galaxy by cosmic rays (about 1 eV per cm^3) is about equal to that contained in galactic magnetic fields, and in the thermal energy of the gas that pervades the space between the stars.

Cosmic Ray Composition

Cosmic rays include essentially all of the elements in the periodic table; about 89% of the nuclei are hydrogen (protons), 10% helium, and about 1% heavier elements. The common heavier elements (such as carbon, oxygen, magnesium, silicon, and iron) are present in about the same relative abundances as in the solar system, but there are important differences in elemental and isotopic composition that provide information on the origin and history of galactic cosmic rays. For example there is a significant overabundance of the rare elements Li, Be, and B produced when heavier cosmic rays such as carbon, nitrogen, and oxygen fragment into lighter nuclei during collisions with the interstellar gas.

The isotope ^{22}Ne is also overabundant, showing that the nucleosynthesis of cosmic rays and solar system material have differed. Electrons constitute about 1% of galactic cosmic rays.

It is not known why electrons are apparently less efficiently accelerated than nuclei.

Cosmic Rays in the Galaxy

Because cosmic rays are electrically charged they are deflected by magnetic fields, and their directions have been randomized, making it impossible to tell where they originated. However, cosmic rays in other regions of the Galaxy can be traced by the electromagnetic radiation they produce.

Supernova remnants such as the Crab Nebula are known to be a source of cosmic rays from the radio synchrotron radiation emitted by cosmic ray electrons spiraling in the magnetic fields of the remnant. In addition, observations of high energy (10 MeV - 1000 MeV) gamma rays resulting from cosmic ray collisions with interstellar gas show that most cosmic rays are confined to the disk of the Galaxy, presumably by its magnetic field. Similar collisions of cosmic ray nuclei produce lighter nuclear fragments, including radioactive isotopes such as ^{10}Be, which has a half-life of 1.6 million years. The measured amount of ^{10}Be in cosmic rays implies that, on average, cosmic rays spend about 10 million years in the Galaxy before escaping into inter-galactic space.

Very High Energy Cosmic Rays

When high energy cosmic rays undergo collisions with atoms of the upper atmosphere, they produce a cascade of "secondary" particles that shower down through the atmosphere to the Earth's surface. Secondary cosmic rays include pions (which quickly decay to produce muons, neutrinos and gamma rays), as well as electrons and positrons produced by muon decay and gamma ray interactions with atmospheric atoms.

The number of particles reaching the Earth's surface is related to the energy of the cosmic ray that struck the upper atmosphere. Cosmic rays with energies beyond 10^{14} eV are studied with large "air shower" arrays of detectors distributed over many square kilometers that sample the particles produced. The frequency of air showers ranges from about 100 per m^2 per year for energies $>10^{15}$ eV to only about 1 per km^2 per century for energies beyond 10^{20} eV. Cosmic ray interaction products such as neutrinos are also studied by large detectors placed deep in underground mines or under water.

Most secondary cosmic rays reaching the Earth's surface are muons, with an average intensity of about 100 per m^2 per second. Although thousands of cosmic rays pass through our bodies every minute, the resulting radiation levels are relatively low, corresponding, at sea level, to only a few percent of the natural background radiation. However, the greater intensity of cosmic rays in outer space is a potential radiation hazard for astronauts, especially when the Sun is active, and interplanetary space may suddenly be filled with solar energetic particles. Cosmic rays are also a hazard to electronic instrumentation in space; impacts of heavily-ionizing cosmic ray nuclei can cause computer memory bits to "flip" or small microcircuits to fail.

Cosmic Rays in the Solar System

Just as cosmic rays are deflected by the magnetic fields in interstellar space, they are also affected by the interplanetary magnetic field embedded in the solar wind (the plasma of ions and electrons blowing from the solar corona at about 400 km/sec), and therefore have difficulty reaching the inner solar system. Spacecraft venturing out towards the boundary of the solar system they have found that the intensity of galactic comic rays increases with distance from the Sun. As solar activity varies over the 11 year solar cycle the intensity of cosmic rays at Earth also varies, in anti-correlation with the sunspot number.

The Sun is also a sporadic source of cosmic ray nuclei and electrons that are accelerated by shock waves traveling through the corona, and by magnetic energy released in solar flares. During such occurrences the intensity of energetic particles in space can increase by a factor of $10^2$to 10^6 for hours to days. Such solar particle events are much more frequent during the active phase of the solar cycle. The maximum energy reached in solar particle events is typically 10 to 100 MeV, occasionally reaching 1 GeV (roughly once a year) to 10 GeV (roughly once a decade). Solar energetic particles can be used to measure the elemental and isotopic composition of the Sun, thereby complementing spectroscopic studies of solar material.

A third component of cosmic rays, comprised of only those elements that are difficult to ionize, including He, N, O, Ne, and Ar, was given the name "anomalous cosmic rays" because of its unusual composition.

Anomalous cosmic rays originate from electrically-neutral interstellar particles that have entered the solar system unaffected by the magnetic field of the solar wind, been ionized, and then accelerated at the shock wave formed when the solar wind slows as a result of plowing into the interstellar gas, presently thought to occur somewhere between 75 and 100 AU from the Sun (one AU is the distance from the Sun to the Earth). Thus, it is possible that the Voyager 1 spacecraft, which should reach 100 AU by 2007, will have the opportunity to observe an example of cosmic ray acceleration directly.

PARTICLES IN COSMIC RAYS

Almost 90% of the cosmic rays which strike the Earth's atmosphere are protons (hydrogen nuclei) and about 9% are alpha particles. Electrons amount to about 1% according to Chaisson & McMillan. There is a small fraction of heavier particles which yield some interesting information. About 0.25% are light elements (lithium, beryllium and boron), but this is greatly enriched over the abundunce of these elements in the universe which is only about one billionth! From this evidence it is implied that these light elements come from the bombardment of heavier particles. Statistical analyses of the light particles point to passage of primary cosmic rays through an amount of matter equivalent to about 4 cm of water.

Medium elements (carbon, nitrogen, oxygen and flourine) are about 10 times their abundance in normal matter and the heavier elements are increased about a hundredfold over normal matter. This suggests an origin of cosmic rays in areas of space with greatly enriched amounts of heavy elements. The density of cosmic rays in interstellar space is estimated to be about $10^{-3}/m^3$.

One interesting aspect of cosmic rays is that they are almost totally matter rather than antimatter. According to Carroll & Ostlie, only about 0.01% of cosmic rays are antimatter, so this sample of the particles of our galaxy provides evidence of the matter-antimatter asymmetry in our galaxy and presumably in the universe as a whole. The few antiparticles that are observed can be accounted for as the results of high energy particle collisions that produce particle-antiparticle pairs.

High energy collisions in the upper atmosphere produce cascades of lighter particles. Pions and kaons are produced, which decay to produce muons. Muons make up more than half of the cosmic radiation at sea level, the remainder being mostly electrons, positrons and photons from cascade events.(Richtmyer)

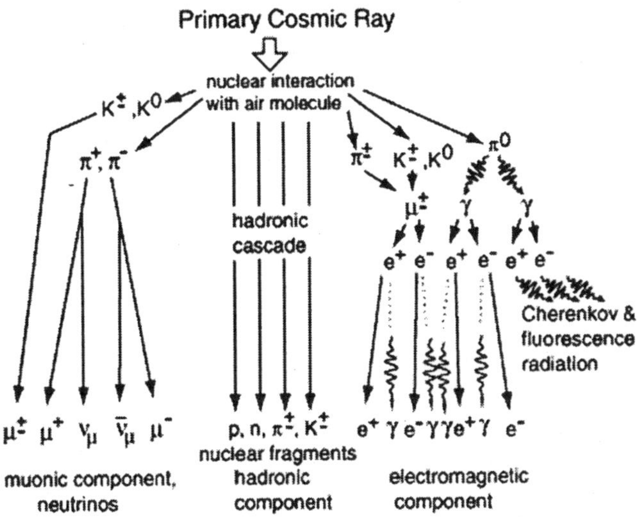

COSMIC RAY SPALLATION

Cosmic ray spallation is a form of naturally occurring nuclear fission and nucleosynthesis. It refers to the formation of elements from the impact of cosmic rays on an object. Cosmic rays are highly energetic charged particles from outside of Earth ranging from stray electrons to alpha particles.

These cause spallation when a cosmic ray (e.g. a proton) impacts with matter, including other cosmic rays. The result of the collision is the expulsion of large members of nucleons (protons and neutrons) from the object hit. This process goes on not only in deep space, but in our upper atmosphere due to the impact of cosmic rays. Cosmic ray spallation produces some light elements such as lithium and boron.

This process was discovered somewhat by accident during the 1970s. Models of big bang nucleosynthesis suggested that the amount of deuterium was too large to be consistent with the expansion rate of the universe and there was therefore great interest in processes that could generate deuterium after the big bang. Cosmic ray spallation was investigated as a possible process to generate deuterium.

As it turned out, spallation could not generate much deuterium, and the excess deuterium in the universe could be explained by assuming the existence of non-baryonic dark matter. However, studies of spallation showed that it could generate lithium and boron. Isotopes of aluminum, beryllium, carbon (carbon-14), chlorine, iodine and neon, are also formed through cosmic ray spallation.

Chapter 8

Modern Electrodynamics

"Action at a distance" and "field of force". Newton's analysis of planetary motions had verified the existence of a gravitational attraction which varies inversely with the square of the distance. The presence of this force could be deduced from Kepler's laws, from the motions of the moon and terrestrial projectiles, and from the mutual perturbations of the planets, which were subjected to detailed investigations by later astronomers and mathematicians. But a further problem remained -the problem of whether a deeper insight into the phenomena of gravitational attraction could be achieved; the problem of somehow understanding and deriving this attraction of gravity from more basic causes.

The fact that Newton had left this question open led many important physicists of his time — notably Huyghens — to mistrust the entire Newtonian thought structure. Without a deeper, more exhaustive explanation Huyghens refused to recognize Newton's formulation of the law of attraction as satisfactory.

At the other extreme, Newton's admirers accepted the law of gravitational attractional as a final, independent world law that required no further explanation. With this attitude they did not, however, express Newton's own feeling; for he had never fundamentally denied the justification of the demands for a firmer establishment of his law. However, he did state emphatically that he would be content with the determination of the presence of this gravitational attraction and the mathematical law for its quantitative description. With this statement he ceased speculating about its origins —

"Hypotheses non fingo". The problem of the cause of the force of gravity, which despite this remained an open problem for a long time, could not lead any further at that time.

But a step ahead was possible with regard to a similar physical law. Coulomb had shown that for attractions and repulsions of electrical charges the force is also inversely proportional to the square of the distance of the mutually interacting charges. Although here, in contrast to the attraction of gravity, a known difference exists — i.e., there are two kinds of electricity, positive and negative, wherein those of similar sign repel and opposites attract — yet, the mathematical law Newton discovered in the attraction of gravity also holds for decrease of force with increasing distance. Physicists of the last century succeeded in obtaining a deeper understanding of these electrical force effects.

How can it be demonstrated that two electrical charges exert forces on each other no matter how far apart they may be? This "action at a distance", this influence of one physical system on another regardless of the distance between them appeared unnatural to the physical intuition of the investigator. The feeling that such a distant effect could not be basic, but required an explanation and a correlation with fundamentals doubtlessly resulted from the influence of materialistic philosophy in the course of historical development; its profound effect on the thought development of western natural science is an unmistakable historical fact. Materialistic philosophy, which taught that atoms of matter, hypothetically introduced by it, were the only seat of action, indicated that the reciprocal effect of atoms through pressure and collision was the only action properly explaining natural phenomena.

Thus it was necessary to explain the apparently independent effects, widely separated from each other, which are stated in Newton's and Coulomb's laws, as results which somehow depended only on pressure and collision of atoms.

Independently of these considerations, which today retain only historical significance, the acute instinct of some investigators of the last century led to the conviction that "action at a distance" must be reduced to a "field of force" effect; that the effects of a physical system could not extend unaided over great distances; that each physical effect must extend continuously in space from place to place.

These convictions were firmly established by Faraday and Maxwell. Faraday, the experimental physicist, allowed his research to be governed by original, new conceptions which in the course of his wonderfully rich experimental discoveries underwent progressive development and verification. These concepts contrasted radically with the ideas of his contemporaries, still entirely built around the principle of "action at a distance". Maxwell was the first to extend Faraday's ideas to include the quantitative wording and mathematical clarity, which they lacked originally.

According to the concepts of this "field of force" theory the reciprocal effects between electrical charges are spread out through the intervening space. Faraday's experiments had shown that these reciprocal force effects were influenced and changed by the presence of dielectrics anywhere within the intermediate space. The idea was firmly established logically that the space between the apparently independent reciprocally acting electrical charges is really a carrier of the potentials and energies distributed in it. These potentials and energies are not directly perceptible to us, but appear in the form of the forces which were described quantitatively by Coulomb's law.

Accordingly vacuum, "empty space", is also a carrier of potentials and energies, the presence of which is not bound up with the presence of matter. Through this consideration Coulomb's law, mathematically expressed, was shown to be an "integral law" which follows mathematically from "differential laws". Coulomb's law appears as the result, as the final effect of more primitive laws which refer to the "field", i.e., to space filled up with electrical potentials. Of course it is true that these laws, simpler to interpret physically, require very much more difficult mathematical tools for their formulation and exact understanding than does Coulomb's law, so simple to express. It must be established here — in a way which emphasizes the principle of continuity — how the state of the electromagnetic field at a given point is related to its immediate neighborhood, and how this electromagnetic state at each point in space and in its immediate neighborhood acts to produce variations in potential with time at that place (time rate of change).

ELECTROMAGNETIC WAVES

It follows logically from the field of force theory that a definite amount of time is necessary for the propagation of electrical effects. The assumption of an absolute distance effect permits the following representation. When an electric charge, which is acting on another far removed charge according to Coulomb's law, is suddenly displaced to another position, then instantaneously, without delay, the other, distant charge will experience the modified effect of the force, which corresponds to the new position.

The latter charge "notices" at once that the position of the former one has been changed. But this idea of an instantaneous propagation of physical effects is incompatible with the idea of a "field of force". If the action of the force exerted by the first charge on the other remote one is effected through the intervening space — reaches the distant charge via the bridge of potentials filling the space between them — then only the field — the bridges — in the immediate vicinity of this charge will be modified to correspond to its new, rapidly changing position. This rearrangement of the field, which commences in the neighborhood of the charge, spreads out into space very rapidly. Nevertheless a finite time is required before this modification extends to the region of the other charge.

The electrical effects of a charge require a finite time for their propagation. If now we consider the case of a charge which not only undergoes a single rapid change in position but is repeatedly and continuously induced to swing back and forth, then the electric field never returns to a static condition. The electric field is transferred into a state of periodic vibration corresponding to the periodic oscillation of the charge — electric waves emanate from the swinging charge. At this point the "action at a distance" and "field of force" theories separate into their experimentally verifiable conclusions.

The mutual actions of static charges can be interpreted both in the sense of "action at a distance" and of "field of force". Despite the essential difference between the methods of representation, problems on static charges always yield the same results. This holds not only for mutual action of static charges according to Coulomb's law but also for many other complex processes in the rich realm of electrical phenomena. The "action at a distance" theory is consistent with Faraday's discoveries, and could be extended in its range of usefulness — with results that agree with experimental facts just as well as the "field of force" theory. It was only the assertions of the "field of force" theory regarding electromagnetic waves, which are said to emanate from oscillatory charges, that made a final decision about "action at a distance" and "field of force" possible, since these waves could not be reconciled in any way with the idea of "action at a distance".

The history of physics contains no more impressive example of the creative force of theoretical thought in physics than the history of the discovery of electromagnetic waves. For these waves, which today rank among the most important problems and media of technical work and which assume a noteworthy position in the general — framework of our modern life, were not discovered through experimental research. They were formulated on paper, educed from the mathematical formulæ which Maxwell had begun to evolve as a quantitative description of Faraday's ideas. Only subsequently were they sought experimentally, and found by Hertz.

A more detailed treatment of Hertzian waves would be superfluous. Originally the discovery of a meditating theoretician, later observed by the experimental investigator, these waves are today familiar to the layman as a result of their technical application in everyday life. Everyone knows that radio waves have wave lengths that extend up to several kilometers. Actually there are electromagnetic waves of much longer wave lengths, but practically these are of little importance. Short wave transmitters utilize wave lengths of about a meter; ultra-short wave (ultra-high frequency) transmitters use wave lengths of the order of magnitude of a centimeter; and experimental results show that infra-red heat rays are simply still shorter electromagnetic waves, whose wave lengths equal only fractions of a millimeter.

LIGHT

Olaf Rümer had determined the velocity of light as early as 1675. In observations that extended over several years he noted the entrance of the innermost of Jupiter's moons into the shadow cast by Jupiter and its emergence from the shadow into the sunlight. If it is correct that light requires a finite time to travel from one place to another then the entrance of one of her moons into Jupiter's shadow will not become visible on the earth until somewhat later.

Likewise a measurable interval of time elapses before the sunlight, begun to be reflected by Jupiter's moon upon its emergence from the shadow, reaches the earth. If this moon, which requires 42 1/2 hours for each revolution around Jupiter, maintained a constant separation from the earth, its emergence from the shadow would become visible to an observer on the earth regularly every 42 1/2 hours.

But while the moon has completed 30 revolutions, for example, the earth has advanced in its orbit so that its separation from Jupiter, and therewith the length of the light path, has been changed. Consequently the observer will see the thirty-first reappearance of the moon not exactly thirty times 42 1/2 hours after the first one, but earlier or later depending on whether the earth and Jupiter have been approaching or receding from each other. From this information Rümer ascertained the value of the velocity of light. Measurements on the velocity of light with purely terrestrial tools did not become possible until much later, when experimental techniques had reached a higher state of development.

Light travels 300,000 kilometers (186,000 miles) per second; it could circle the equator almost eight times in a second; but it reaches the earth about eight minutes after leaving the sun, and requires about four and one-half years to travel to the earth from the nearest fixed star. It takes 100,000 years for light to traverse the milky way system rectilinearly. And astronomers have knowledge of systems related to the milky way, located far beyond it, from which light travels 100 million years before reaching us; thus we see these "nebulae" as they were 100 million years ago.

The mere fact that light has the same velocity of propagation as the electromagnetic waves predicted by Maxwell would not have sufficed to justify the proposition that light was a special case of electromagnetic radiation. But light had already been described as a wave radiation in Maxwell's times, and more refined optical experiments (concerning the "polarization" of light) indicated properties of light which conformed exactly to the law of radiation mathematically constructed by Maxwell.

Spectral resolution of light and the synthesis of white light from colored had already been observed by Newton in investigations which are no less worthy of admiration than his celestial mechanics.

But Newton did not consider light — like sound -to be a wave motion. He preferred the interpretation that a luminous body, such as the sun, constantly emits hail showers of very tiny particles and that this shower of light corpuscles is the basis for what we perceive as light. Another explanation, already worked out by Huyghens in Newton's time, was suggested by later research but was not experimentally confirmed until afterwards, although Newton himself had described experiments which could only be understood with the aid of this new idea.

If we imagine two hail showers falling on the same place simultaneously, or two machine guns bombarding the same target, the result is necessarily an intensification, an addition of their effects. In the case of light, however, there are cases where light coming together with light produces darkness. This phenomenon — called "interference" — was investigated in innumerable varied experiments; upon its practical application are dependent many of the most important observation instruments of modern physicists (e.g., spectroscopic "gratings").

The existence of these interference phenomena becomes understandable if one observes the intersection of the circles (waves) produced on the water's surface by two stones thrown into it. The circles continue through each other without being destroyed; but where the wave crest of one and the wave trough of the other come together, they annul each other. That is the basis of the interference in which two light excitations annul each other.

These interference phenomena as we see them show that light is the result of a wave process. All interference effects ever observed, including their finest details, can be explained quantitatively by the conception of light as a wave motion, wherein, strictly speaking, monochromatic light always corresponds to a fixed wave length. Red light with a wave length of about .00076 mm. and violet at about .00038 mm. represent respectively the longest and shortest wave lengths visible.

In this physical picture of light processes there are no more colors; there are only oscillations in a colorless electromagntic field. But we shall guard against the careless methods of expression that found favour in an earlier stage of natural scientific development and philosophical adjustment. We want to guard against a mode of expression such as the following: now the "existence" of light is disclosed and the coloration of light is recognized as an illusion — as something which originates in our brain. As we know, no less a figure than Goethe spoke out passionately against Newton's theses on the colour of light.

Certainly the establishment of the wave properties of light has brought us no closer to the real existence of blue or red, although refined experiments have taught us of the existence of remarkable new properties of blue and red light. Just as when an object which has always been seen from the front is viewed from the rear for the first time; the new things that are learned are added to what was already known; but it would be erroneous to say that through this consideration of the rear side the "existence" of the entire object has been recognized and its front side has been proven an "illusion".

We shall still defer the questions which arise here and which in general concern the most profound problems of physical knowledge and thought. We shall still have sufficient opportunity to reconsider the questions of natural scientific doctrines.

THE ETHER PROBLEM

This discussion, in striving to clarify the thought content of modern physics and to make its philosophical conclusions evident, is not bound to the historical sequence of development. Our attempt to follow logical relationships leads us along a zigzag path through historical development.

Long before light was recognized (Maxwell) as a special case of electromagnetic radiation, the problem of how light waves could pass through a vacuum had been pondered. Sound waves, in fact, produce vibrations of the atmosphere and cease to exist in the absence of air. But how can we explain that light oscillations can take place also in a vacuum? What oscillates here? To satisfy these questions a new hypothesis filled empty space with an obliging medium, fine, thin and all-pervading — the ether. Then the problem arose of inferring the nature of this ether from knowledge of the laws of light phenomena and conversely, explaining the laws' of light by a mechanical model of the ether.

When these questions arose at the beginning of the last century the tools for their answers were already at hand in the highly developed mechanics of that time. Based on the mechanics of centers of mass, a mechanical theory of a material medium which fills all space was set up. The theory of elastic vibrations and waves in gases, liquids and solids had essentially been established. Here we must differentiate between longitudinal and transverse waves; i.e., respectively between those in which the matter oscillates parallel to the direction of propagation and those in which the vibration is perpendicular to this direction. Gaseous media sustain only longitudinal waves, in which case each minute volume of the medium is periodically rarefied and compressed. Both types of waves occur in solid bodies.

Light waves had already been proved — through the aforementioned experiments on the polarization of light — to be transverse, never longitudinal. Thus if the ether is to be considered as a material medium, it must be conceived of as a sort of solid body which opposes any rarefaction or compression with infinitely great resistance (possesses no "compressibility"). This result was surprising in view of the fact that the planets move on through this ether without encountering the least resistance.

But if many other difficulties had not arisen this fact alone would not have proved decisive. In any case, not a single tangible new idea, not one suggestion for an experiment leading to new conclusions resulted from all the laborious experiments on the mechanical ether model.

Maxwell's classification of light phenomena with in the comprehensive framework of electromagnetic processes produced a completely altered situation. There resulted not only a knowledge of entirely new relationships between light and other physical phenomena and regularities but a changed spiritual attitude toward these problems was facilitated. Maxwell had succeeded in establishing the regularity of electromagnetic waves — inclusive of light — with a completeness and mathematical clarity which do not suffer by comparison to Newton's law of gravitation. As Newton had, Maxwell could say, "Hypotheses non fingo".

Through Newton's law all questions concerning the motion of planets, moons, comets, etc., became answerable without waiting for a possibly more thorough establishment and explanation of this law. Similarly, through Maxwell's law all questions falling within the region of validity of his theory could be answered precisely and independently of the problem of the mechanical explanation of the properties of ether.

Maxwell himself did not follow the principle "Hypotheses non fingo" absolutely; he had deliberated over various extensions for obtaining a mechanical ether theory. But the overall total direction of his work pointed to something entirely different. It was the influence of his discoveries which led more and more physicists to regard the entire problem of the mechanical ether model as fruitless and superfluous.

For if it is possible to predict the result of any conceivable electrodynamic experiment, if electromagnetic processes can be grasped as readily mathematically as the motions in the planetary system — then what more do we want?

Not only the method of thinking trained by materialistic philosophy but also that physical instinct, which had learned from nature the more profound significance of the physical "field of force" laws as against the "action at a distance" laws, resisted the recognition of Newton's (or similar to this, Coulomb's) law as a non-derivable physical elementary law. There remained only one counter argument in which traditional materialistic philosophy opposed Maxwell's theory, which already depended on the idea of a "field of force" and had clarified this idea with mathematical precision. This philosophy maintained that only mechanical laws, as evidenced in the pressure and collisions of atoms, are to be recognized as the true and final laws of nature, and that therefore only the reduction of natural phenomena to a mechanical model can yield a real "explanation", a real "understanding" and a real "recognition" of the existence of things.

Since Maxwell's theory appeared as a mere description of electromagnetic processes without really attempting to explain their laws, it must have startled the advocates of the mechanical explanation of nature when Kirchoff in 1876 began his lectures on mechanics with the famous sentence, that it was the function of mechanics "to describe the motions proceeding in nature completely and in the simplest manner." In that sentence he expressed a trend of thought which had already been amply explained and confirmed previously by Ernst Mach and which is usually designated as physical "Positivism".

This reference, which showed that the ultimate aim of mechanics is simply a description of mechanical events — that it does not in any way lead us to any "understanding" of the "essence" of mechanical phenomena — stripped the defence of the mechanical meaning of the ether of its last weapon. When it became evident that belief in a separate appreciation of mechanical laws actually depends on habit only, all inducement for considering the mechanical laws as fundamental, like the Maxwellian electrodynamics, disappeared.

The long since decided victory of Maxwell's theory over the mechanical ether theory — conversely, modern physics traces mechanics back to electrodynamics — is therefore bound up with an impressive displacement of the philosophical attitude of physical investigation. The new approach can be characterized as a radical answer to the old.

It relinquishes just that which previously was considered — be it in a vague sense or in the sense of the materialistic philosophy — as the true goal of physical research; namely the penetration into the heart of natural physical processes by stripping them of all the cloaks of outer appearance. In contrast we set a much more unassuming goal. We recognize that we can achieve our final purpose only by obtaining new data through experiments and by winning support for the prediction of the results of future experiments by a mathematical description of the experimental determination developed by the acuteness of theoreticians.

THE RELATIVITY PRINCIPLE

The existence of a luminiferous ether filling the empty interstellar space would make it still possible to speak in a certain sense of "absolute" motion; namely, motion relative to the fixed ether.

Then the statement that light travels 300,000 kilometers per second would mean that light moves forth with this velocity relative to the fixed ether. In this case it must be possible to determine an absolute motion through opticalelectromagnetic experiments, and we have already seen that the detection of an absolute motion is impossible through purely mechanical experiments.

This problem was thoroughly investigated experimentally. The most famous example is the experiment conducted by Michelson. He measured the possible difference in light velocities on the earth in mutually perpendicular directions with very great accuracy; and the resultant difference was exactly zero. Obviously this experiment suffices to prove that there is nothing to be gained from the primitive idea of a fixed ether permeating interstellar space. All kinds of ways out were attempted, for example, the hypothesis that the ether surrounding the earth was carried along with the earth in its motion. But such an idea leads to very great difficulties. How far into interstellar space does this zone of ether carried along by the earth extend? How much ether does the sun carry along with it?

Maxwell pointed out that if the notion of a permanently fixed, space-filling ether were correct the absolute motion of the solar system must be detectable through observations on Jupiter's moons, in an extension of Römer's investigations. But experience denies this, since the sun is really moving relative to the fixed star heaven.

Should it therefore be assumed that as the earth carries along its surrounding ether, the entire planetary system also draws along a self-enveloping ether cloud? Certainly the concept is becoming hopelessly complicated; and the many very accurate experiments conducted along these lines must lead only to a maze of most complicated results. That is not the case. The Michelson experiment and a series of further experiments which we shall touch on later in part yielded conclusions which in themselves are very simple and clear. The problem of developing an incontrovertible, logical connection between the varied (in themselves simple) experimental facts requires rather deep penetration into the lowest strata of our physical representation structure, which had to be revolutionized and transformed to agree with these facts.

"Aberration", a phenomenon discovered by Bradley in 1727, provided very simple, impressive counter-evidence against the "carrying along theory". If we stand on the front platform of a moving train in vertically falling rain we get wet despite the presence of roofing overhead. For the raindrops falling perpendicular to the fixed earth (aside from all the secondary influences of the air motion produced by the moving train) are falling obliquely relative to the train. An analogous phenomenon is evident in the light of the stars falling on the earth, which leads to the conclusion that in the course of a year the fixed stars apparently execute small elliptical motions in the sky relative to the earth's circular motion.

Finally, let us consider the Doppler effect. When a whistling locomative is approaching us we hear a higher tone than when it is receding from us. This is so because more sound waves sweep past our ear per second as the whistle approaches than when the locomotive is standing still or receding. An analogous phenomenon manifests itself in the case of light. From the displacements which the spectral lines in the spectra of the fixed stars undergo due to this Doppler effect, both the earth's annual motion and the motions of various fixed stars towards and away from us can be inferred.

In the case of sound waves the magnitude of the Doppler effect is not only dependent on the relative motion of the sound source and the listener, but on the motion of both relative to the atmosphere. Correspondingly, if there were a fixed ether, one should be able to recognize absolute motions from the optical Doppler effect. Actual experience shows that this is not the case, that the optical Doppler effect depends solely on relative motion and furnishes no possibility of determining an absolute fixed ether. We can not summarize here the abundant detail of corresponding experimental facts which bear out the relativity principle, not only in mechanics, but in all branches of physics.

But even with this determination — which deals the ether representation a death blow — we are still far from understanding the problem; it is now really posed for the first time.

An attempt to describe our complete optical-electromagnetic knowledge coherently in the all-inclusive form of the principle of relativity encounters tremendous difficulties. To surmount these we must revise the usual, apparently indispensable concepts which to us are most self-evident. For though it is simple to say that the relativity principle is also valid in optics, we must accustom ourselves to the idea that light always has the same velocity relative to any moving body — therefore, that the assertion "that light travels 300,000 kilometers per second" not only holds true for a particular condition of motion of the observer but for all conditions of his uniform rectilinear motion.

But how can this "principle of the invariance of the velocity of light" be established free from contradiction when elementary thought habits assume that if we speed ahead in a space ship at 40 kilometers per second behind a light ray the velocity of the light ray relative to us would appear diminished by the 40 kilometers per second?

THE THEORY OF RELATIVITY

It required all the acumen of the greatest physical thinkers of our time to solve this puzzle. Modern publicity frequently supports the opinion that the so-called "theory of relativity", in which these questions are cleared up, is a quite personal, private discovery of the famous physicist, Albert Einstein. Whereupon they usually conclude that the challenging position taken by the Third Reich toward Einstein personally with respect to political views must necessarily result in a challenge of the theory of relativity.

This is a misunderstanding. It should be noted that a number of other investigators also produced definite contributions to the theory of relativity (Poincaré, Lorentz, Minkowski, Planck, Hilbert, Weyl, Eddington, etc.) and further, that the physical knowledge expressed in the theory of relativity would have been an inescapable logical conclusion from the experimental facts if Einstein had never lived. An attempt at a systematic explanation of the theory of relativity would exceed the bounds of this treatment. Just a few points are singled out to help us examine the philosophical attitude of modern physicists more thoroughly.

If we look out of the window of a train traveling 80 kilometers per hour at another train traveling in the opposite direction with the same velocity, we can not determine our train's velocity relative to the earth by seeing only the other train.

The only thing we can measure is our velocity relative to the second train — which relative velocity is obviously 160 kilometers per hour. What is the basis for our conviction that this is the case — a conviction of which we are so certain that not a single person will bother to actually determine it by measurement? We possess very deeply rooted thinking habits that impel us to this opinion so forcefully that we are wont to consider a divergent result as merely logically foolish. As is known, Kant developed the clever idea that our concept of nature is conditioned essentially not through inherent qualities of the objects of nature but through changeless thought patterns not derived from experience, but incorporated by us into observations and investigations of nature. He not only expressed these ideas in a general form; but he also analyzed these ways of perception completely, attributing them invariably to the nature of the human mind.

The physical experiences which lead to the theory of relativity exhort us to a fundamental distrust of constructions which denote certain experimental data as independent and invariant. For the theory of relativity teaches us to recognize that a number of just such measurements are capable of (and require!) modification and generalization.

Previously there existed the inclination to believe in the certainty of their accuracy independently of experimental experience. Let's take the above example of the two trains again: today we know that the proposition that the velocity of the two trains relative to each other is equal to the sum of their (oppositely directed) velocities relative to the earth is not really precisely correct. Actually this relative velocity is a little smaller than the sum of the two.

In this example the difference is imperceptibly small; but, if instead of the railroad trains. we take two bodies which are moving with much greater velocities quite a considerable difference results. If each of the two trains were traveling at a speed of 200,000 kilometers per second (twothirds of the velocity of light) the relative velocity would only amount to 276,923 kilometers per second instead of 400,000.

There is experimental evidence to confirm this. The former opinion that appeared certain without experiment was disproved by investigation which confirmed it as practically correct only under certain conditions (namely, only for velocities that are much less than that of light). This extension and change of ideas, that appeared to us as invariably self-evident, and the development of new concepts, proven necessary by experimental experience, made possible the very great usefulness of the theory of relativity.

In passing we noted the "positivist" attitude toward the problem of physical knowledge — merely seeking a summarizing description of experimental facts instead of an alleged explanation, penetrating into the essence of things. This makes it necessary to check most rigorously whether all our assertions, suppositions and problems fit completely into a system of pure description of observed results. Any statement which falls beyond this limit — especially any attempt to express something about the so-called "essence" of physical things — must be eliminated and declared basically meaningless.

It is overrating the significance and scope of physical knowledge to believe it possible to make statements that are not confined to a description of experimental measurements — may these measurements be confirmed, suspected, questioned or contradicted. Modern developments have shown over and over again with finality that only this radical liberation of "senseless" opinions brought about by positivist criticism gives scientific thought the necessary freedom to adapt its ideas to the greater and more difficult demands placed on it; demands due to the high precision applied to the experiences of daily life by modern experimental research and to the disclosure of new, separate fields which lie completely outside of our previous experience and thus completely beyond the power of comprehension of our customary, acquired modes of viewing. Conceptions which appear most self-evident to us are frequently just the ones which prove to be the greatest obstacles to the transformation of our ideas necessary to adapt them to the new experiences.

The theory of relativity includes one of the most wonderful examples of the liberating force of this positivist criticism on our opinions. What is meant by the phrase, two events occur "simultaneously"? No problem exists when both events take place in spatial proximity, and can thus be seen in the same glance. But what is meant by the assertion that the falling of a stone on the earth occurs "simultaneously" with an event on Sirius? The key to the theory of relativity lies in the insight that, to make any sense at all, the definition of the "simultaneity" of two events widely separated in space must be made concrete, must be transposed into an experimentally verifiable form of expression.

No method but the following remains to effect this definition. At the instant the stone falls on the earth a light signal is emitted; at the same time when the imagined event is taking place on Sirius a light signal is sent out from there; and when the earthly signal reaches Sirius, a light signal is sent back from there at once. The reflected light signal returns to the earth again about eighteen years later. If the time elapsed between the earthly event and the arrival of the first light signal from Sirius amounts to nine years (more exactly, if it is just as great as the separation between the times of arrival of the first and second light signals from Sirius) we say that the two events, on Sirius and the earth, occurred "simultaneously".

If signals existed that were propagated more rapidly than light signals these would naturally be used for the definition of simultaneity. If the theory of "action at a distance" were correct the difficulties of defining simultaneity would disappear in general. For then it would be possible (in principle — the question of the technical accomplishment is an independent one) at the instant the stone fell on the earth to dispatch a signal which would become perceptible on Sirius without any loss of time at all — perhaps through the Coulomb type of "action at a distance". But we have accepted the bases of the "field of force" theory, according to which all physical effects require a finite time of propagation — an idea of decisive meaning for the theory of relativity. By the "field of force" theory the velocity of the signal can not exceed a definite limit; and today we know that light possesses the greatest signal velocity possible.

After the above definition for the notion of simultaneity has been given, mathematical investigation leads to the perception of simultaneity as a relative concept. Viewed from the abovementioned train, which is traveling at 200,000 kilometers per second, the two events on the earth and Sirius would no longer take place simultaneously, since an experimenter traveling in this rapidly moving train would obtain a different result in executing the measurements pertaining to the definition of simultaneity than we would on the earth.

The remarkable results which ensue. Consider a space ship traveling with an enormous velocity — almost equal to that of light. Suppose that the crew returns to the earth after a one year voyage at a speed slightly less than that of light. [1] Their watches, taken along in the space ship, have measured just one year of time; their one year stock of provisions has just been used up; and their hair has greyed just about as much as would be expected from the hardships of a one-year trip in interstellar space. But, arrived on the earth the crew finds that in the meantime the human race has aged one hundred years.

These are very remarkable assertions; and our thinking which finds it too difficult to leave familiar paths, is inclined, at first, to view them as pure nonsense. But that is prejudiced. The sum of these propositions forms a self-contained, incontrovertible logical system; a system which does not stem from the imagination, but is based on the irrefutable facts of experimental experience.

Finally, it should also be mentioned that in connection with the theory of relativity — in a generalization of theory which extends far beyond what we have discussed until now — the problem of tracing Newton's gravitational attraction and "field of force" laws back to fundamentals was solved. The fact that the "gravity mass" of any body exactly equals its "inertia mass" is conclusive for this solution.

Otherwise expressed, in ideal fall (in a vacuum) all terrestrial bodies experience exactly the same acceleration. Thus one can say that a physicist in a high elevator which is released and falls down unhindered will make exactly the same observations as another physicist who moved uniformly, in a straight line through the universe in a space ship in a region where the gravitational attractions of the stars are ineffective because of their great distance. To the physicist in the freely falling elevator the weight of any body would be annulled; and the physicist in the space ship, when his motion was accelerated, would find the same effects in his ship as the earthly gravitational field exerts.

This consideration leads to a more penetrating understanding of Newtonian gravitation. We must deny ourselves more exact treatment of this. One thing, though, deserves emphasis in connection with these considerations: the complete validity of Euclidian geometry in actual physical space must be denied; only approximate validity should be ascribed to it, since more general geometric notions, which we owe to the German mathematician Riemann, agree more closely with real physical space (Einstein). This also shows that we obstruct the path of physical knowledge when we consider opinions and methods of perception, which appear self-evident from long familiarity, as irrefutable, invariable and independent of physical experience.

Chapter 9

Reality of Atoms

SPECULATIVE ATOMISM

Actual proof of the reality of atoms is an achievement of this century; even at its beginning critical consideration required the confession that atoms, although investigators spoke of them, were still no more than a stimulating and useful hypothesis, which in the final analysis was unproved and possibly unnecessary, perhaps even misleading. Although the idea of atomism was important and fruitful at the very beginning of western natural research, this idea did not originate from western research itself; just as we borrowed the model for mathematical thinking from the Greeks, we took the atomistic interpretation of nature from them.

Nothing exists — Democritus taught — except atoms and vacuum; everything else is imagination. These innumerable atoms, indestructible and invariable elementary constituents, are to be considered the basis for all beings and events in nature. Individual atoms possess invariable geometric form and motions which fluctuate due to the pressure of and collision with other atoms. In all changes in the structure of nature the atoms are always preserved; from nothing comes nothing; nothing which exists can be destroyed; change is merely combination and separation of parts.

The differences of all things stem from the diversity of their atoms in number, magnitude, structure and arrangement. Since the movements of each individual atom are regulated according to natural law, nothing arbitrary can happen in all nature; nothing happens accidentally, but everything from a cause and of necessity. Our rough senses can not recognize these amazingly fine elements in their true nature and form, but experience their effects only vaguely. "Only in the imagination does sweetness exist, in the imagination bitterness, in the imagination heat, cold, colors; in reality nothing exists but atoms and vacuum."

This is marked evidence of that "decoloration" of the world, the assurance that all direct sensory observation is an illusion and that a knowledge of the true status of things must lead us to a picture in which nothing remains but the geometrical form and mechanical motion of the atoms.

These idea forms of materialistic philosophy introduce the possibility of considering all events in nature as results of strict regularity; the gist of the basic concept of natural scientific thought is anticipated here in general and is thoroughly formed for the first time. On the other hand we must not forget to recognize how radically this philosophy opposes religious representations of its time. For the Greeks every stream was a God; in the springs lived the nymphs, in the woods the horned Pan, and in caves and caverns resided demons.

This entire mythological picture of the world, which suspected the arbitrary and incalculable influence of demoniacal powers in every natural process, was pushed aside by the powerful conception of a picture of nature of stricter regularity. Epicurus, who renewed Democritus' teaching, and who in his thinking and conduct of life combined a rationalistic, vigorous attitude with a very cultivated, fine-spirited being, was not so brutal as to deny existence completely to the Gods.

He let them continue as blessed, immortal beings, occupied with themselves, and never infringing upon the workings of the world — these workings being developed according to the mechanical lawfulness of atomic motions in strict, definite sequence. In a great didactic poem *"De rerum natura"* the Roman Lucretius explained completely the ideas of atomistic regularity for all the fields of nature known at that time. These teachings were made known to western scholars in connection with the humanistic studies of the Renaissance;

Gassendi, principally, introduced Greek philosophy into western science.

Let us consider the ideas which Newton, together with Gassendi, applying the divergent theories of Descartes, formed about the construction of matter. F. Dannemann explains them as follows in his historical work [1] on the natural sciences: "He considered it most plausible, that it (matter) consists of solid, impermeable, moving particles.

Since natural bodies, e.g., water, are invariable in their properties, the particles of which they consist must neither be able to be used up nor destroyed. The variation of material things is to be laid exclusively to the separations, combinations and movements of those invariable particles." These are, as is obvious, exactly the concepts of materialistic philosophy, almost unchanged, just as Democritus stated them. But Newton was — outside of natural science — in no wise an adherent of materialistic theories. For him these ideas meant nothing but a clue to physical research; he did not see in them the support of radical, world-viewing results.

EVALUATION OF THE ATOMIC CONCEPT

Dalton first recognized and effectively utilized the stimulating force of the atomic concept and its ability to guide future physical research. To him we owe the use of the idea of atoms for understanding basic regularities of chemistry. Chemistry makes a sharp distinction between mixtures of different substances and chemical combinations. A mixture — as, for example, a solution of sugar in water or a mixture of nitrogen and oxygen — can include arbitrary (within certain limits) relative proportions of the constituents. In a nitrogen-oxygen mixture the portion of oxygen can be varied continuously by addition or removal; correspondingly, the mixture exhibits properties somewhere between those of pure oxygen and of pure nitrogen, depending upon the relative concentration.

Also, the mixture can be separated into its constituents through relatively superficial means. In a chemical combination entirely new properties appear, which can be quite dissimilar to those of the constituents. The proportions of the components of a chemical compound are quite stable and cannot be varied continuously.

Thus exactly 16 grams of oxygen always combine with 2.016 grams of hydrogen to form water. Another chemical substance, hydrogen peroxide, can be formed from 2.016 grams of hydrogen and just 32 grams of oxygen. It is striking that exactly twice as much oxygen is utilized in the second case as in the first; and on the basis of the atomic concept Dalton found a clear explanation for such facts, which occur similarly in all chemical combinations. His explanation was as follows: The "molecule" of water, i.e., the smallest possible particle of water, contains only one atom of oxygen; the molecule of hydrogen peroxide however contains two atoms of oxygen, while it contains just as many (two) atoms of hydrogen as the water molecule.

Here we arrive at the distinction between atoms and molecules; we designate the smallest particles of chemical compounds as molecules, which for their part consist of atoms of the chemical elements. According to Dalton's ways of thinking a comprehensive study of proportions by weight in chemical compounds rendered the determination of the relative atomic weights of all the elements possible. The "atomic" weight of oxygen is arbitrarily designated by the value 16, whereupon hydrogen acquires the atomic weight 1.008 and every other element, similarly, receives a definite atomic weight derived from chemical weight relationships.

The regularities which the gases exhibit, which in the case of sufficiently small densities assume an "ideal" form, proved very helpful for the clarification of these ideas; for example, in the sense that a gas mass contracts to one-half its original volume when, with constant temperature, the pressure impressed upon it is doubled.

Criteria existed for the hypothesis that not only in compounds but also in a gaseous element such as nitrogen the particles moving at random through space are not necessarily identical with atoms; in nitrogen, for example, each "molecule" of the gas consists of two atoms. In the chemical combination of gases (with constant pressure and temperature) there is a simple relation between proportions by weight and volume relations (proportions by volume) between the chemical compound and the still uncombined constituents.

To fit these facts Avogadro introduced the explanation that equal volumes of all gases measured at constant pressure and temperature always contain the same number of molecules. Based on this "Avogadro's principle" a "molecular weight" can be determined for each gas independent of chemical weight relations.

The molecular weight of a chemically homogeneous gas is equal to the mass of 22.4 liters of this gas measured in grams at one atmosphere of pressure and a temperature of 0° C. Experience shows that values obtained in this way correspond with the chemically defined atomic weights. In many elements, e.g., the metals, this molecular weight is exactly the same as the atomic weight; thus metallic vapors are "monatomic". In the aforementioned nitrogen, however, the molecular weight is exactly double the atomic weight. And in water it equals 18.016 (2 X 1.008 + 16) in accordance with its above-described chemical composition.

These ideas are further verified by simple relations of molecular weights such as, "diffusion velocities", or by the characteristic difference between the specific heats of "monatomic" and "polyatomic" gases. None of these, it must be emphasized, are proofs for the correctness of the atomic concept; but they are proofs for its usefulness. The atomic idea describes very plainly a large number of important and comprehensive regularities, thus fadlitating greatly our progress with these phenomena.

Crystallography furnishes evidence of further regularities which are certainly clearly explained by the atomic idea, though they too do not lead to proofs for the reality of the atom. In the crystalline state of matter the atomic concept suggests a regular grouping of the atoms or molecules, alongside of each other and arranged in layers. This view induces important results; on this basis the variety of possible forms of crystals are shown mathematically to be limited very considerably. The completion of this mathematical proposition yields the wonderful result that according to the atomic concept there should be no more nor less than 32 different "crystal classes" which are defined by different properties of symmetry.

Actually this confirms precisely the experience of mineralogists; there really occur in nature all the crystal forms (crystal symmetries) which are compatible with the atom idea, but not one single one which contradicts it. The so-called "law of rational indices" discovered by the mineralogists formed an important adjunct to this original consideration, which is related only to the symmetry properties of different crystal forms. If we imagine the crystal as built up of a regular stratification of layers of atoms, then its outer boundary surfaces could never lie so that they partially cut through the atoms. The limits for the possible positions of the boundary surfaces given by this view are exactly equivalent to the requirements which the law of rational indices places on the position of the crystal surfaces.

Such experiences must necessarily encourage the more and more energetic pursuit of the atomistic notion — which intrinsically was so clear — even though tangible proof for the reality of atoms was still lacking. Actually the theoretical development of the atomic concept was furthered extensively by the most able physicists. We have already mentioned ideal gases and indicated how the gaseous state of aggregation should be represented on the basis of the atomic theory; the molecules of a gas are completely separate from each other; each one moves along with high velocity until it meets another one, and then, in elastic rebound, their direction of motion and speed change.

The continuous impact of countless molecules on the walls of the enclosing vessel produces the pressure of the gas, perceptible to us with our rough tools. These ideas were elaborated thoroughly (Clausius, Maxwell) and our understanding of the gaseous state of aggregation was developed very clearly and completely. Whereby the proof resulted that Avogadro's principle, first only speculatively suspected, must necessarily be correct if the atomic hypothesis proves at all valid.

The discovery of the energy principle (Mayer, Joule) further strengthened confidence in atomism. It was recognized that mechanical energy, which disappears through friction or the like, reappears in the quantity of heat produced — in such a way that a definite amount of consumed mechanical energy corresponds to a definite number of calories of heat. Conversely, according to the same conversion relation, in steam engines quantities of heat are transformed again into mechanical kinetic energy.

This conversion of mechanical energy into heat and its converse can be considered quite consistent, in the sense of "macrophysics". Whereupon one must establish that energy, which has proved indestructible, can assume both the form of a quantity of heat and of mechanical energy. The "macrophysical", "phenomenologic" heat theory thus arrived at is entirely sufficient for answering all the questions encountered in connection with heat engines, and with heat conversion in chemical processes.

But further thought is suggested to give the process of the conversion of mechanical energy into heat a clear interpretation in the light of the atomic theory. In a gas, as was represented above, the concepts of heat and temperature are not applicable to the individual molecules, but only to the gas mass as a whole. In an individual molecule only the kinetic energy, i.e., its velocity of motion, or at best also its rotation, can be changed.

Thus, if energy is added to a gas mass, which appears macrophysically as added heat, it can only mean that the average energy of motion of the gas molecules has increased. The exact determination of the connection between the average energy of motion of the gas molecules and the temperature of the gas is a part of the theoretical investigations on the "kinetic theory of gases", which has already been discussed. Similarly, in a solid body, a crystal perhaps, the temperature and heat energy contained within it are conceived of as mechanical energy of its molecules or atoms. Although in their regular arrangement in layers no one atom can leave its place, fine vibrating motions within the crystal — possibly similar to the fine vibration of a heavily traveled steel bridge — are still possible. If these vibrations within the crystal are so small that they are not perceptible to us as mechanical motion, we do note the energy of these motions — as heat content which manifests itself in the temperature of the body. (If the atoms contained in each small piece of the crystal move by very small amounts in random directions, the whole crystal appears motionless to our macrophysical senses.)

Apropos of such thought processes was Boltzmann's atomistic interpretation of a remarkable regularity in heat phenomena. If we imagine that a planet's motion is interrupted and the planet then is induced to move in the opposite direction with the same velocity, it will retrace its entire elliptical path in the opposite direction after the reversal. This is an example of the fact that all purely mechanical motions are, as we say, reversible. The upward motion of a stone, thrown from the earth, (in the ideal case) until its-point of reversal is an exact temporal mirror image of the subsequent downward motion. Similar reversibility — which can also be denoted as a physical "symmetry of the positive and negative time orientation" — exists in all purely electromagnetic properties.

But when we move a body over a table surface against frictional forces and heat is generated thereby, there is no reason to expect that in moving the body backwards along the same line the converse is true — that the heat energy will change back to energy of mechanical motion. Or take an electric current which flows through a wire and produces heat — when it flows through the wire in the opposite direction it again produces heat; the heat energy does not change back into electrical. These are examples of irreversible processes. Another example of such a process is the mixing of two liquids; in most cases these intermix by themselves, but this process never runs backward of itself. A further example of importance to us is the following: if a gas filled vessel is introduced into an evacuated one and is opened there, the gas diffuses throughout the entire available volume. It is not impossible to restore the original conditions; the larger vessel can be evacuated again by pumping and the gas can be compressed into the smaller one.

But this does not constitute an exact reversal — a temporal reflected image — of the first process; and it can be considered that for the restoration of original conditions after any irreversible process a definite "compensation" must be included (Clausius). In the steam engine, which undertakes to convert quantities of heat into mechanical energy, there does not occur a simple reversal of the process of the production of heat through friction, but each steam engine functions according to a scheme similar to the following: a certain amount of heat is removed from a heat container maintained at a high temperature.

A fraction of this amount is changed into mechanical work, while the remainder is conducted into a heat reservoir of lower temperature. As "compensation", therefore, for the conversion of heat into work there occurs the transfer of a quantity of heat from a hotter body to a colder one — i.e., therefore a process which nature could perform "irreversibly" itself, since in heat conduction the heat always flows from higher to lower temperature.

The occurrence of such "irreversible" processes among heat phenomena posed a very difficult problem for the "kinetic theory of heat" — tracing back the laws of heat to the mechanics of atoms. For purely mechanical processes were shown to be always reversible, which makes it appear impossible to trace these irreversible heat phenomena back to mechanical properties.

The solution of this difficulty, achieved through Boltzmann's perspicacity, runs as follows: in an aggregation of many similar particles, such as are attributed to every physical system by the atomic theory, there is no point to pursuing the motion of each individual atom or molecule; it is only important to us to observe the average statistical behaviour of large numbers of atoms. Thus statistical concepts are necessarily introduced into the consideration and we must determine which (defined in a rough statistical sense) events are to be considered as especially probable (thus as occurring very frequently) in comparison with other conceivable processes.

For our example ths signifies that when the small gas filled vessel is opened in the large empty one "it is extremely probable" that the gas will stream out and distribute itself uniformly in the large vessel. Strictly speaking it cannot be definitely known that this will happen; indeed, to view the processes of motion in the gas mass with the same complete certainty achieved for planetary motions one must know quite exactly the position and velocity of each individual molecule at the beginning of the experiment. The fact that, instead of this, we simply denoted the initial condition of the experiment statistically stipulates that we can not predict its further progress with complete certainty, but only with very great probability.

The reversibility of pure physical processes, which according to the atomistic conception is the basis for macrophysical heat phenomena, depends upon the existence of an exact reversal, an exact temporal reflected image for every process — whereby it can happen that the gas mass distributed in the abovementioned large vessel contracts into the smaller one, in an exact reversal of the normal process. No one can guarantee that this is completely out of the question; but we can show that such an event is extremely improbable — mathematically its probability is expressed by an infinitesimally small number. Thus it becomes clear that despite the fundamental reversibility of atomistic elementary-processes, yet, on a large scale, in macrophysical events, practically irreversible phenomena take place.

Another instructive example of an irreversible process is the mixing of two gases or liquids with each other. According to the kinetic-atomic theory this problem is somewhat similar to the one wherein two types of balls — red and white ones for instance — are thoroughly shaken up in a sack. If the red and white spheres are carefully separated initially, they will become completely mixed up after further agitation. If, then, we continue to shake, it is possible in principle that the original separation of red and white balls will be restored — but practically this possibility is of little moment since it would require shaking for astronomically long periods of time before this "spontaneous separation" could be expected with appreciable probability.

A quantitative measure for the irreversibility of a thermodynamic process (from a purely macrophysical standpoint) can be specified in the form of "entropy" — a quantity which always increases to a maximum in irreversible processes and does not decrease again (Clausius). According to Boltzmann this entropy can now be defined through the atomic concept as a measure of the disorder in an atomic aggregate.

THE LIMITS OF THE DIVISIBILITY OF MATTER

Everyday experience, which appears to show unlimited divisibility of matter, through simple refinements makes it certain that atoms must be extraordinarily small, and that therefore the number of atoms in a gram of any substance must be enormously large. For an exact measure of this quantity we use the so-called Loschmidt number. As a "gram-atomic weight" of an element we denote a mass of a number of grams equal to the atomic weight; the Loschmidt number is the (equal for all elements) number of atoms in a gram-atomic weight. Evidence for the divisibility of matter can be obtained by distributing a minute amount of a strong-scented substance (mercaptan) in a large quantity of air and then determining to what degree of dilution the odor can still be perceived; or by dissolving a strongly colored liquid (eosin) in a relatively large quantity of water and then noticing to what degree of dilution a uniform coloration of the water remains perceptible.

A still more suitable experiment involves the use of a very dilute solution of fluorescein, a substance which fluoresces strongly in incident light. Extremely small volumes of the solution are examined with a microscope to determine to what degree of dilution a spatially uniform fluorescence can be detected. From such observations it can be inferred (Perrin) that the Loschmidt number is definitely greater than 10^{21} (a 1 with 21 zeros after it).

But quite simple, daily experiences occur that indicate the limits which its atomistic structure places on the subdivision of matter. These involve utilization of very thin membranes of matter. Gold leaf, which is used for gilding purposes, is hammered down to an exceptional thinness — when held against the light it exhibits a greenish translucence — of about 100 millimicrons (a million millimicrons make a millimeter); and yet this thickness is still far removed from the ultimate limit. But there exist direct indications for the atomistic constitution of matter in commonplace, familiar soap bubbles, often formed when washing by a film of soapy water stretched between the thumb and index finger.

These bubbles iridesce in variegated colors, whose diversity and rapid shifting is evidence of the variations in the thickness of the bubble with position and time. Small dark spots appear in a thus colored bubble, which at first may be considered as holes, but which in reality are enclosed by still thinner soap films. If bubbles like this are produced in a solid frame instead of in the hand, and if they are protected from evaporation in a vapour-filled enclosure, they can be preserved for days. Newton had disclosed that inside these black spots — i.e., this membrane which weakly reflects light — still blacker, thus still thinner membranes are formed.

The thickness of the thinnest membrane obtained by Newton was about 6 millimicrons, about 20 times thinner than gold leaf. But the next thicker membrane obtainable, just before this thinnest one, is of exactly double thickness. This provided a tangible indication of the molecular structure of this skin substance; obviously, the thinnest membrane represents a single layer of molecules, while in the next thinnest one two layers are piled together. A very thin film can be produced more conveniently by allowing a minute amount of oil to spread out on a large water surface.

The familiar iridescent films which oil forms on water correspond in their thickness approximately to the usual soap bubble. Much thinner oil films which are no longer directly visible but can be recognized with simple tools can easily be formed. In these thicknesses of about 1 millimicron have been reached; and in these it has again been shown that the thicknesses of the finest films are not continuously variable. As in the soap bubbles, we find a quite definite thickness prescribed for the thinnest, second thinnest, etc. oil film (for a definite previously determined type of oil). The limits of the divisibility of matter have actually been reached here.

Further highly informative investigations were performed on very small corpuscular particles. Here the invention of the ultra-microscope rendered important service; it made visible not only the form, but also the position and motion of particles which cannot be "seen" in the usual sense, since there are limits, defined by the nature of light, for the optical observation of very small objects. Bodies of dimensions smaller than the wavelength of visible light cannot be "formed" in any way by light rays; cannot be made visible in their true form. The ultra-microscope, through an ingenious device, made it possible to detect optically the presence of particles which are somewhat smaller than the wavelength of visible light.

For the fundamental task — to prove the reality of atoms experimentally — two types of investigations in particular have rendered real contributions. If very small particles are placed in a liquid it appears that they do not sink rectilinearly to the bottom and remain there but rather they move through the liquid in an irregular, erratic manner. The motion becomes more intense the higher the temperature of the liquid is raised; but it is independent of a minor agitation of the enclosing vessel or of other accidental disturbances. Also, if left undisturbed, the erratic motion of the particles continues for days or years.

This Brownian movement serves as a direct proof for the correctness of the proposition, set up by the atomic theory, that the heat content of a body is to be conceived of as an internal motion which is so fine that it cannot be macrophysically detected as such. One must not conclude that there are very fine motions or currents in the liquid which produce the Brownian movement of the suspended particles.

That would lead to the idea that the motions of a throng of closely neighboring particles were somewhat similar as is the case with dust we see around us in the sun's rays. In Brownian movements two approaching particles in a liquid are completely independent of each other, and thus separate again very soon. This shows that the hidden motions in the liquid, the permanent presence of which must be regarded as characteristic of the heat condition of the liquid, are completely "irregular" and macrophysically absolutely imperceptible.

Brownian movements have become the subject of many experimental investigations, for theoretical considerations had shown (Smoluchowski, Einstein) that its precise observation permits definite conclusions about the motions of atoms and molecules, through irregular collisions with which the particles in Brownian movements are driven around. It was possible to infer from investigation of Brownian movements how great the average kinetic energy of the individual molecules must be. And since the total heat energy contained in the liquid is equal to the sum of the energies of the individual molecules, we can infer how many molecules there are in a given volume. This leads to a determination of Loschmidt's number.

Related investigations and considerations are also feasible in many kinds of "fluctuation phenomena". Imagine an armor plate suspended from chains and blown upon by a very strong, uniformly operating sand blast apparatus; the plate will be forced from its equilibrium position by the pressure of the blast and will then hang in a slightly different position. Now, we remove the sand blast apparatus and by bombarding the plate with machine guns exert the same total pressure on the middle of the plate; the armor, struck over and over again by the single shots again assumes the displaced equilibrium position, this time continuously oscillating irregularly about this equilibrium position.

Observing just these oscillations, one can determine the strength of the single blows striking the plate. Similarly in many types of physical apparatus very fine irregular fluctuations can be observed and from them the magnitude of the Loschmidt number can be determined. In the very fine physical apparatus the presence of the irregular heat motion of the atoms can be detected; thereby most varied methods can be utilized for determining Loschmidt's number. The values of Loschmidt's number obtained from all such investigations have always agreed within the limits of their accuracy.

The following describes the second possible way of determining Loschmidt's number from investigations on very small particles: the density of the earth's atmosphere decreases, as we know, with increasing height. Imagine two very high cylinders — about 100 kilometers high — to be filled respectively with two different gases, perhaps oxygen and nitrogen. It is demonstrated that in the gas whose molecules are heavier (oxygen) the density decreases outward from the earth's surface more rapidly than in the other gas.

The gas molecules are hindered by their kinetic energy from simply accumulating on the bottom of the vessel, i.e., forming a liquid. But the heavier they are, the stronger is the effect of gravity forcing them downwards and thus the more rapidly does the gas density decrease with increasing height above the earth's surface. The mathematical law relating to this states that the comparison of the decrease in density in the two cylinders depends solely on their molecular weights.

This was most useful in its application to the production of an artificial gas in which the actual mass of the individual molecules is known. Perrin laboriously produced large quantities of small resin pellets of equal weight. He obtained these resin particles from an emulsion which originally contained particles of the most varied magnitudes; but by an ingenious procedure, involving painstaking work, he was able to separate out particles of corresponding mass whose diameter was from 200 to 300 millimicrons.

These particles were then placed in water. (We know that with sufficient degree of dilution dissolved substances, e.g., sugar in water, behave analogously to the ideal gases.) Thus, the resin particles placed in the water represented an ideal gas and vessels a kilometre high were no longer needed to determine the decrease in density with increasing height with this artificial gas; even in small containers these artificial gas molecules, enormously heavy in comparison with usual molecules, crowd together noticeably at the bottom. Through measurement of this effect and comparison with the known decrease of atmospheric density with increasing altitude it is possible to compare the directly determined mass of these artificial, huge molecules with the masses, absolute values still unknown, of the chemical molecules present in the atmosphere. Thus a new approach to the weighing of molecules — or differently expressed, to a determination of Loschmidt's number — is reached.

The numerical value which resulted from these different determinations of Loschmidt's number (and from further determinations still to be discussed) is 6 x 1023 (a 6 with 23 zeros after it). If the hydrogen atoms contained in 100 grams of water were distributed over the entire earth's surface, one atom would fall on each square centimeter. . Finally we come to the experiments in which single atoms are actually isloated and made, in a manner of speaking, tangible, so that there is no longer any possible doubt concerning their real existence.

Let us begin with Laue's famous discovery of crystal interference. Modern workers in physical optics and spectroscopy no longer use only prisms for the spectral resolution of light; the diffraction grating has become much more important. When numerous lines at equal, close intervals are scribed upon a polished surface, a monochromatic (including just one wave length) light ray will neither be reflected from this surface as from a usual mirror nor as from a rough surface which reflects diffusely toward all sides. The ray is reflected in a series of definite directions which vary according to its wave length; in all other directions there is no reflection because the light excitations traveling in these other directions mutually annul each other by interference.

This knowledge (a special interference experiment) gives us, as previously mentioned, the clearest and most obvious proof for the wave character of light and makes it possible to determine the wave length of the light just from, a knowledge of the spacing of the grating lines. (In common gratings this is between a hundredth and a thousandth of a millimeter.) Conversely naturally the separation of the rulings on the grating, if by chance this value is not known, can be determined from the reflection produced by this grating with light of known wave length.

X-rays, on traversing crystals, exhibit remarkable interference effects which are related to those of the "line grating" described but are considerably more complicated. The discovery of these effects proved two things with one stroke; first, that X-rays are a wave radiation; and second, that crystals possess a fine grating-like structure, which is so fine that it defied observation with former tools.

Careful analysis of the varied and complicated interference phenomena which can be obtained in this way led to the certainty that this internal fine structure of crystals is exactly the same as that which atomistic concepts led us to expect; the "illumination" of crystals with X-rays clearly displays their synthesis from atoms and thereby conclusively assures the reality of these atoms.

The wave lengths of X-rays are much shorter than those of ordinary light (X-rays, like light, are also included within the general bounds of the classification of Maxwell-Hertz electromagnetic waves). Because of their short wave lengths it is not possible without further refinement to attain diffraction (interference) of X-rays with the usual, relatively rough optical line gratings.

But in crystals nature presented us with "natural diffraction gratings" with which we can study X-ray interference effects. With a very refined device it was possible to obtain diffraction of X-rays with an optical diffraction grating (line grating) so that the wave lengths of X-rays could be measured just as optical wave lengths are. If we know the wave length of an X-ray, from the interference effects which result from the passage of this X-ray through a crystal we can infer the volume relations within the crystal. X-ray interference effects in crystals not only substantiate the reality of atoms with indubitable clearness and distinctness, but also facilitate another determination of the number of atoms in a macroscopic piece of matter — a new determination of Loschmidt's number.

Millikan (in extending and refining older, less lucid ones) performed an experiment which was striking in the simplicity of its fundamental idea. In it he provided tangible proof for the atomic nature of electricity. A mist of minute oil drop lets is sprayed into a chamber; many of these are electrically charged in the process of their production.

One droplet is singled out and observed through a microscope. Left to itself it falls vertically downward, due to the force of gravity; not according to the ideal laws of free fall, but rather with constant velocity, since air resistance is proportionately more effective against these small droplets than against larger bodies. If the droplet is subjected to the influence of an electric field, of proper intensity perpendicular to its direction of fall and opposing it, the droplet will start to rise, with a constant velocity, the magnitude of which is dependent upon the field strength.

The ratio of the forces working in both cases and the magnitude of the charge carried by the droplet can be calculated from a comparison of the speeds of rising and falling. Consideration of the values obtained for numerous such droplets indicates that the magnitude of the charge does not vary continuously from case to case; none of them ever has a smaller charge than the so-called "elementary charge", and larger values are integral multiples (2, 3, 4 . . .) of this smallest one. Thus we have tangible evidence that electric charges cannot be artibitrarily divided indefinitely; rather, all electric charges are composed of indivisible "elementary charges", each of which carries a charge of 4.77×10^{-10} (4.77 divided by a 1 with 10 zeros after it) electrostatic units.

This atomism of electricity is closely related to the atomism of matter in general. A law concerning electrolysis, discovered by Faraday and named after him, states this relation. Helmholtz had early concluded from this law that if the concept of the atomism of matter were proven correct an atomism of electricty must be assumed. If one zinc and one copper plate are immersed in an aqueous solution of copper sulfate and an electric current is passed through the solution in a positive direction from the zinc to the copper plate, the zinc gradually becomes dissolved in the liquid as further copper is deposited on the copper plate.

Quantitatively this experiment shows that when one gram atomic weight of zinc has been dissolved, just one gram atomic weight of copper has been deposited; and corresponding weight relations are found to hold for all similar electrolytic experiments. Moreover, the number of gram atomic weights deposited or dissolved always remains in a simple proportion to the quantity of electric charge which has passed. Thus there occur here regularities of the same character as the laws of weight relations in chemical reactions which Dalton explained by the concept of atomism. In these laws Dalton saw a criterion for the definition and use of the atomic concept; by the same right we can, with Helmholtz, infer the atomistic structure of electricity on the basis of Faraday's law.

Conversely, knowing from Millikan's work that the atomism of electricity not only expresses an auxiliary idea but is a real fact, we can infer from Faraday's law that the atomic structure of matter is a reality (which has already been verified in our consideration of crystal interference). The knowledge of the magnitude of the elementary electric charges furnishes one more basis for calculation of Loschmidt's number, characteristic of this atomistic structure of matter.

This in no way exhausts the experiments which show the indubitable reality of atoms. Further direct proofs for atomism are gained in connection with the investigation of radioactivity. From the observation of radioactive preparations it was determined that "alpha-radiation" is simply an emission of electrically charged helium. When this alpha-radiation strikes a suitably prepared zinc sulfide screen, even with limited intensity, it produces a series of small light flashes (scintillations) there.

From this effect it is seen that the emitted electrically charged helium is not an arbitrarily divisible, indefinitely diffuse substance, but must consist of discrete particles. By counting the light flashes, it became possible to determine how many single atoms constitute a definite quantity of helium; a new method of determining Loschmidt's number. The results again confirmed those obtained by other methods; this confirmation is evidence that each of these small light flashes actually does come from a single (electrically charged) helium atom.

Of even greater significance was still another process, in which again the effects of single atoms (or "ions", as electrically charged atoms are usually referred to) became visible. A moisture saturated atmosphere is produced in a carefully cleaned, dust-free vessel. If the size of the container is suddenly increased by means of a moving piston the saturated atmosphere cools off and becomem "super saturated" — the water vapour begins to condense to liquid droplets. But the first droplets require "condensation nuclei" on which to form, and by removing all very fine particles (dust, etc.) we have eliminated the usual nuclei.

If an alpharadiation from a radioactive preparation strikes the atmosphere of this "Wilson Cloud Chamber", each of the rapidly traveling helium ions produces a fine track, along which are present innumerable atoms or molecules, now likewise electrically charged, which were hit by the alpha particles shooting through the air. Along the entire path these newly produced ions form suitable condensation nuclei; the entire path appears marked as a fine streak of mist visible to the naked eye. Again the effect of a single charged atom — and this time still more beautifully and definitely than by scintillations -has become visible, and again calculations from such "cloud tracks" yield Loschmidt's number.

The Wilson chamber has become one of the most important research tools of modern atomic physicists. Obviously its usefulness is limited to very rapidly flying particles, since only these possess sufficient energy to "ionise" innumerable other atoms or molecules along long paths. But in the processes connected with radioactivity these particles appear under various circumstances; and the Wilson chamber presents a beautiful opportunity to study all the details of the processes associated with their formation and transformation.

Geiger's development, the counting tube, is an equally important tool for atomic physics — for determining effects of single, isolated atomic particles. As soon as a single rapidly moving electrically charged atom enters this apparatus a macrophysically perceptible electric charge is produced by an ingenious arrangement. Like the Wilson chamber this apparatus makes it possible to establish the reality of atoms, to determine Loschmidt's number and to study thoroughly atomic physical processes.

CONTRIBUTIONS OF ATOMIC PHYSICS

Before Millikan's determination of the elementary electric charge, electricity — notably through investigations by Lenard had been represented in pure culture, so to speak, namely in the form of cathode rays formed in a highly evacuated electric discharge tube — also in an X-ray tube, for example, where they produce X-rays by impinging on a solid plate. These cathode rays consist of a flow of negative electricity; from the results of Millikan's experiment it is certain that this electricity must flow in the form of separate, indivisibly small particles. These particles have been named "electrons"; they might also be designated as the atoms of electricity.

The cathode ray which travels rectilinearly when undisturbed can be forced into a curved path through the use of electric and magnetic fields. The motion of an electron deflected under the influence of known electric or magnetic fields must be calculable by Newton's second law — force equals mass times acceleration — if we know the ratio of its charge to its mass, since the acting force is proportional to the charge on a single electron.

Conversely, this relation of charge to mass can be inferred from the experimental determination of the deflection of cathode rays in electric and magnetic fields.

Since the charge on an electron is known from Millikan's experiment (naturally from the outset it is assumed highly probably that the electrons in a cathode ray possess a single Millikan elementary charge, not two or three; further experiments confirm this as fact) the mass of an electron can be calculated from the measurements on cathode rays. The result shows that the mass of an electron is about 1838 times smaller than that of a hydrogen atom; or, expressed in grams, equals 0.9 x 10-27 (0.9 divided by a 1 with 27 zeros after it) grams.

The method of determining the ratio of charge to mass with cathode rays can also be extended to electrically charged atoms, ions. We know that from hydrogen atoms only one type of ion can be formed — a particle which has almost the same mass as the hydrogen atom itself and possesses exactly one positive elementary charge. Two different forms of ions — each of almost the same mass as the atom — can be formed from the helium atom; one has a single positive elementary charge, the other has two. The impression results that the electrically neutral hydrogen atom contains positive and negative charges within its structure. Obviously hydrogen contains just a single positive and a single negative electric charge; and the negative one must be connected with a proportionally minute mass, whereas almost the entire mass of the

Conversely, this relation of charge to mass can be inferred from the experimental determination of the deflection of cathode rays in electric and magnetic fields.

Since the charge on an electron is known from Millikan's experiment (naturally from the outset it is assumed highly probably that the electrons in a cathode ray possess a single Millikan elementary charge, not two or three; further experiments confirm this as fact) the mass of an electron can be calculated from the measurements on cathode rays. The result shows that the mass of an electron is about 1838 times smaller than that of a hydrogen atom; or, expressed in grams, equals 0.9 x 10-27 (0.9 divided by a 1 with 27 zeros after it) grams.

The method of determining the ratio of charge to mass with cathode rays can also be extended to electrically charged atoms, ions. We know that from hydrogen atoms only one type of ion can be formed — a particle which has almost the same mass as the hydrogen atom itself and possesses exactly one positive elementary charge. Two different forms of ions — each of almost the same mass as the atom — can be formed from the helium atom; one has a single positive elementary charge, the other has two. The impression results that the electrically neutral hydrogen atom contains positive and negative charges within its structure.

Obviously hydrogen contains just a single positive and a single negative electric charge; and the negative one must be connected with a proportionally minute mass, whereas almost the entire mass of the hydrogen atom is attached to the positive one. The supposition is suggested — and abundant experience raises it to certainty — that the negative charge in the hydrogen atom is just a negative electron. The positive hydrogen ion, from which no further electrons can be separated, is designated as the "nucleus" of the hydrogen atom; it is also referred to as a "proton". In the helium atom two negative electrons are needed to neutralize the positive charge of the "nucleus"; here again ionization of the atom means simply the removal of one or both electrons. The nucleus of the helium atom is, moreover, identical with the aforementioned alpha particles.

As Rutherford recognized, the structure of all atoms is similar to these. In each atom almost the entire mass is concentrated in a positively charged nucleus; the effect of the "envelope" or "cloud" of negative electrons surrounding this nucleus is to render the atom electrically neutral. Thus the number of these electrons must always be equal to the "nuclear charge number"; i.e., equal to the number of positive elementary charges in the heavy nucleus.

Chemists have known for a long time that a very convenient view of the chemical elements can be obtained through their arrangement in the "periodic table of the elements" -wherein the elements are arranged (Meyer and Mendeleef) according to their atomic weights. It was later shown that the atomic weight is not the characteristic by which the various chemical elements are differentiated that atoms of different atomic weights can belong to the same element (isotopes) and that atoms of equal atomic weights can belong to different elements (isobars). Not the mass of the nucleus, but its charge really determines the chemical nature of an atom;

for the atomic number (nuclear charge number) determines the structure of the electron shells since it dictates how many electrons the neutral atom must possess. Through the structure of the electron shells the chemical properties are determined. Chemical molecules are formed when two or more atoms combine; in the combination the outer rings of electron shells experience certain transformations, but the nuclei remain unchanged.

The size of an atom can be determined in many ways. Their diameters are all equal to about 0.1 millicron. These diameters cannot be specified by exact numbers like the masses can, since the atoms are certainly not smooth spheres with definitely defined surfaces. As in a cloud with blurred boundaries, in an atom only an approximate value can be specified for its diameter. Such values can be determined from crystals, in which the atoms must be packed together very closely.

From investigations of gases it is possible to obtain corresponding determinations of these atomic magnitudes. The atomic nuclei are very much smaller; perhaps a hundred thousand times smaller in diameter than the atoms. Thus, almost the entire volume of an atom is occupied by the electron shells. But an electron is essentially no larger than an atomic nucleus. How it is possible that despite this, in a hydrogen atom for example where the entire "shell" consists of only one electron, the volume of the entire atom is "filled" up by this electron is a question to which we must return.

While chemical processes involve only the outer electron shells, there are other processes in which the nuclei themselves experience transformation, division or synthesis. In the main these are the already noted processes of radioactivity; but also numerous artificial, arbitrarily produced transmutations have been produced in atomic nuclei in recent years. Thus the nuclei must also be represented as complex structures; today we trace all nuclei back to two building stones, one of which is the already described proton and the other the "neutron" — a particle which has nearly the same mass as a proton but is electrically neutral.

Chapter 10

The Paradoxes of Quantum Phenomena

LIGHT QUANTA

The Greek philosophers of the materialistic school -even though differences exist in the manner of notation and expression; for example, we do not use the word "atom" today for an indivisible structure, but rather for one which is built up of still smaller parts.

We compare, not what we call "atoms", but electrons, protons and neutrons, with the atoms of the Greeks. There is irony in this development, since the discovery of these new, confusing facts preceded the real proof for the reality of atoms. In 1900 Max Planck made the discovery that directed atomic research along new paths and opened to the theoretical and experimental work of physicists a tremendous new realm for research. Thus the most important developments for atomic physics were crowded together at the turn of the century: and at that time there appeared the criticism, based on Positivism, of atomistic ideas developed until then.

First Ernst Mach made it clear that in reality the atomistic hypothesis was still completely without proof and totally unnecessary at that time. However, shortly after 1900 the first experimental proofs for the reality of atoms were accomplished. In 1900 Planck had already made the discovery whose real meaning was grasped only a few years ago and from which we gradually realized that atoms are entirely different from what old philosophers thought they were.

The wave theory of light is distinguished from the "corpuscular theory" of light embraced by Newton — whereby light is said to consist of a shower of very fine particles, "corpuscles" — by the proposition that the energy of light is subdivided continuously at pleasure. This proposition appeared to be confirmed by primitive experience, as had that of the unlimited divisibility of matter.

In the region of the planet Neptune the same amount of light energy which illuminates one square centimeter on the earth's surface has spread out over a surface of 900 square centimeters; arrived in the vicinity of Sirius this light energy has spread out so far in space that it illuminates about 10,000 square kilometers of surface, yet within the limits of the differentiation ability of the human eye this distribution is still quite uniform, as is seen inversely in the radiation coming to us from Sirius.

The sparkle of the fixed stars is due to disturbing effects in the earth's atmosphere; an observer on the moon would see these stars shine with regular brilliancy. The wave theory of light, according to which light is propagated continuously through space, maintains that this expansion of light energy can be continued indefinitely.

But there are experimental results which contradict this idea. Let us consider the "photoelectric-effect", the regularities (Lenard, Einstein) of which can be explained briefly as follows: if ultra violet light of sufficiently short wave length impinges on a metal surface, a constant emission of electrons from the illuminated metal takes place.

The remarkable thing is that with a decrease in light intensity the number of electrons emitted per second decreases correspondingly but the velocities of the individual electrons remain constant. If the wave theory of light is correct, we must conclude that the electrons torn loose from. the metal can possess less energy, the less intense the light is — just as branches are torn off with less force, the weaker the wind blows. But actually, with constant wave length, no more rapid electrons are produced by the strongest light intensities than by the weakest.

This might still be understandable if the single electrons gathered energy from the light waves until they had received a definite amount before shooting out. But then a finite time would have to elapse for very weak light intensity before the emission of the electrons could begin. This is also not the case. Remarkable experiments were performed; a low intensity light source was set up so that it would take hours for a single electron to obtain the amount of energy the emitted electrons actually possess.

Even under these con ditions not even the smallest measurable interval of a second elapsed between the beginning of the illumination and the beginning of the electron emission. We do not acknowledge that something completely unintelligible for the wave theory of light has occurred here; yet it appears that Newton was somehow right, and that the incident ultra-violet light is to be compared to a hail shower, its energy being concentrated in innumerable corpuscular particles.

It was also possible to determine the exact law of this concentration of energy from the photo effect. A monochromatic ray of light (which according to the wave theory includes just one wave length) by this theory merely possesses equal "light quanta", each of which contains an amount of energy that is proportional to the wave length; or, otherwise stated, proportional to the frequency (number of light vibrations per second) of vibrations.

The ratio of light quantum energy to frequency of vibration is the same for all light quanta; it is designated by the letter h (Planck); this famous quantity which controls all of modern physics — it is called the Planck quantum of action — has the value of 6.6 x 10-27 (6.6 divided by a 1 with 27 zeros after it) erg-seconds and the dimension of "action" (energy times time).

Newton of course was not aware of this connection between the energy of light corpuscles and frequencys — and he could not have found it, since it is not a statement which fits into a light theory which appears internally incontrovertible, but rather, presents a connection between two different, diametrically opposed theories of light; the wave theory on one hand and the corpuscular on the other.

The presence of these light quanta in radiation can also be recognized in the radiation itself; not in as tangible, lucid a form but as the result of more profound considerations and experiences. Consider a black walled vacuum vessel; if the walls are held at constant temperature the vessel will be filled up with electromagnetic radiation. For these black walls constantly emit a definite radiation — only heat radiation at lower temperatures, but also visible light at higher temperatures, as is seen in a glowing iron.

The radiation constantly emitted by the walls of the vessel is absorbed again upon reaching the bounding wall, but a certain amount of radiation energy will remain present permanently. The problem, upon solution of which Planck discovered his quantum of action, is this: what is the intensity distribution in the radiation (usually called "black body radiation") which fills this hollow volume? What proportions of infra-red, red, yellow, violet, ultra-violet radiation are contained therein and how is their intensity divided among the various wave lengths of the spectrum when this radiation is resolved spectroscopically?

This problem had already been attacked experimentally and theoretically by many investigators before Planck. The theoretical investigation had led to a remarkable obstacle. It had been shown that the "kinetic" theory of heat answered this problem definitely; if the most fundamental principles of physics are correct and the kinetic interpretation of heat phenomena is admissible, a definite law must hold for the spectral resolution of black body radiation. This law was called the Rayleigh-Jeans radiation law. Its theoretical derivation depends upon considerations which make it possible to compare the average energies of the different light waves in a black cavity with the average energies of single atoms in gases or solids.

The Rayleigh-Jeans law was actually confirmed for the infra-red side of the black body radiation spectrum. (More exactly, these are the facts: the Rayleigh-Jeans radiation law is valid at a fixed temperature of black body radiation — i.e., of the container walls — for a part of the spectrum, from the longest wave lengths down to and including part of the region of shorter wave lengths.

The higher the temperature becomes, the further the range of validity of the Rayleigh-Jeans law is extended toward the shorter wave lengths.) From this experimental confirmation of the RayleighJeans law on the infra-red side of the black body radiation spectrum another possibility was presented for the determination of Loschmidt's number; for from measurements on black body radiation it can be inferred how great the average kinetic energy of the atoms must be at certain temperatures.

Actually, however, the Rayleigh-Jeans law is not absolutely valid; the further we go from the longer toward the shorter wave lengths the less exactly this law is fulfilled. At very short wave lengths the difference becomes serious. The Rayleigh-Jeans law produced an absolutely erroneous result for very short wave lengths; according to it the short wave lengths must possess so much energy that the total black body radiation summed up from very large wave lengths to infinitely small ones becomes infinitely great.

If this is to be. taken as striking proof for the falseness of the Rayleigh-Jeans law (in relation to short wavelengths), Jeans' expedient — there is no real black body radiation; cavity radiation can only be partially black (with respect to the longer waves) — must be considered. But then the black walls would send radiation energy into the cavity unlimitedly; the energy of the spectral region within which black body radiation is already present would continue to increase with time and more and more radiation would be piled up in the short wave length region. These conclusions contradict our experience and the physical theories which lead to the Rayleigh-Jeans law must necessarily be in error. This is of serious import for physical theory because no special hypotheses had been required for this law, now proven to be false. A revision is necessary in the most primitive, fundamental assumptions and ideas of our physical theory.

Planck, who discovered the true law of black body radiation was forced to new and curious considerations to be able to interpret the law he had formulated.

Essentially his new ideas — although at first he sought to formulate them with as much caution and restraint as possible — meant nothing else than a radical break with the wave theory of light; the light quanta produced experimentally by the photoelectric-effect were also necessary on the basis of Planck's observations.

The concept of light quanta solves the difficulties of the Rayleigh-Jeans law quite simply; a constant increase of intensity at the short wave length end of the spectrum is prevented by attributing high energy to the light quanta corresponding to short wave lengths (high vibration frequencies).

Investigation of the Compton effect gave important verification of the light quantum theory. According to the Maxwell theory of light, when a light wave spreads over free, unbound electrons the following must result: the electrons are induced to periodic oscillation through the periodically varying electrical forces of the light waves. In turn, each of these oscillating electrons will then — like a small radio transmitting antenna -emit new electromagnetic radiation in all directions.

This is referred to as a "scattering" of the radiation incident upon the electrons; part of the energy of the primary radiation is deflected from the original direction of propagation and is divided in all directions around the "scattering centre". When this process is investigated experimentally by the use of light with very rich energy quanta (for the effect to be noticeable, X-ray "light" must be used) the results fall beyond the bounds of the electromagnetic wave theory of light and again clearly demonstrate the presence of light quanta.

An in crease in wave length of the scattered light is established. (According to the above explained proposition, the scattered radiation should have exactly the same frequency of vibration as the primary radiation since the frequency of the scattered radiation is determined by the frequency of the "antenna", of the electron, and the electron oscillates in the same rhythm as the primary wave exciting it.) In its quantitative regularity the increase in wave length of scattered radiation (discovered by Compton) permits recognition of the physical process which limits it.

The validity of the inferences concerning this was later verified by further experiments utilizing the more refined methods of the Wilson cloud chamber, the counting tube, etc. Today we are certain that collisions of light quanta and electrons are the basis for the Compton effect. The single process of this phenomenon appears as follows: a single light quantum impinges against an electron. The electron is displaced in its motion by the collision and the light quantum is deflected from its original direction of flight. The collision proceeds according to the laws of elastic impact; i.e., the total energy of both bodies after impact is the same as before impact; the "centre of gravity principle" or the principle "action equals reaction" is also valid.

This latter principle (called the centre of gravity principle or the principle of conservation of momentum) is as important for modern physics as the principle of the conservation of energy. Like the energy principle, the momentum principle was first recognized and formulated in mechanics. The de signation "centre of gravity principle" stems from the following description of its effect: an acceleration of the common centre of gravity of several bodies can never result from mechanical reciprocal action of these bodies; e.g., with all the mutual actions of the sun and planets the centre of gravity of the system (which almost, but not exactly, coincides with the centre of gravity of the sun) always retains its uniform rectilinear motion relative to the fixed star heaven.

Or when a cannon is shot, the same force is exerted on the cannon as on the projectile, but in the opposite direction. From this formulation (equivalent to the other one) there results the expression "action equals reaction". Moreover the relativity theory, the results of which are indispensable for the quantitative understanding of the regularities of the Compton effect, demonstrated that this principle of the conservation of momentum is intimately related to the energy principle.

Thus the collision of light quantum and electron proceeds according to the laws of conservation of momentum and energy. If the direction of flight of a single one of the scattered light quanta is known, the direction of motion of the electron hit by this quantum can be calculated exactly. It has been confirmed experimentally that for each electron struck one light quantum is scattered (Compton-Simon, Bothe-Geiger).

QUANTUM TRANSITIONS

In the observation of a non-luminous Bunsen flame containing a little sodium vapour (common salt introduced into the flame) through a prism or diffraction grating, the spectrum shows only a single line[1] — in the yellow; the entire visible portion of the light includes only the one wave length corresponding to this line. Conversely, if light, shown by spectroscopic analysis to contain all visible wavelengths is passed through sodium vapour the only light absorbed is that corresponding to this yellow spectral line; the sodium vapour transmits all other visible light. This famous discovery of Kirchhoff and Bunsen became the foundation of "spectrum analysis".

Like sodium, every other neutral element possesses characteristic spectral lines — in the visible region, however, not a single one like sodium, but usually a large number of lines; e.g., iron has several thousand lines in the visible spectral range alone, and for each element there are additional lines in the infrared and the ultra-violet. Small displacements of spectral lines and resolution of a single line into several closely adjacent ones can be obtained by exposing the light-producing atoms to the influence of electric or magnetic fields (Stark effect, Zeeman effect). But with this exception the spectral lines of each element are rigidly defined; therefore, the elements present in the flame can be determined from the investigation of the spectrum of a flame.

For chemists this has become one of the most important methods of detecting very minute traces of elements. Light sources in which molecules rather than atoms produce the light have shown that each type of molecule possesses a definite characteristic spectrum. Spectral analysis became a most important, fruitful aid to astronomy.

The spectra of stars and nebulae contain chiefly the same spectral lines which we can produce by introducing different elements into terrestrial light sources. Fundamentally this furnishes proof that the entire stellar world is composed of the same chemical elements we have become familiar with on our earth. Furthermore, investigation of stellar spectra yielded an abundance of knowledge concerning the physical nature of the different heavenly bodies and the conditions to which luminous atoms are exposed therein. Particularly, several lines were found in stellar spectra which — as an exception to the rule — could not be reproduced in any laboratory on the earth and furnished evidence that matter exists there under conditions which cannot be imitated in the laboratory (e.g., gases of absolutely tremendous rarefaction).

Even in these cases there appear no chemical elements other than those known from earthly experience. Spectra furnished more refined and more abundant information on the nature of atoms, than any other method of investigation. It seems apparent that there is more to be learned about the nature of the iron atom from the spectrum with its many thousand lines than from the little numerical data concerning mass, atomic number and approximate size of the atom. But first it was necessary to understand the language of these spectra, and that was not at all easy.

That inner motions of the atoms were betrayed on the outside through light emission was not surprising in itself; for it is known that the constituents of the atom are electrically charged and in motion must act like a small antenna. But it was not possible to interpret the regularities of the spectra through an explanation of the radiation concept based on the principles of classical mechanics and electrodynamics. Careful analysis demonstrated simple and beautiful mathematical regularities in the spectra. The spectral lines of hydrogen, for example, can be represented in wonderfully simple mathematical formulae (Balmer, Lyman, Paschen); and Ritz formulated a quite general mathematical law (named after him) of very simple form for all elements. But it was shown to be impossible to obtain a physical explanation of these empirically determined regularities on the basis of the physical concepts established before 1900. Despite its important contributions to chemistry and astronomy all spectral investigation remained completely outside the theoretical structure of physics of that time. The regularities found by Balmer, Ritz and others, which were not self-explanatory, remained with all of spectroscopy in the museum of physical science.

This condition was altered by the development of the quantum theory from Planck's discovery, first hesitantly and in later years more and more rapidly and forcefully. In 1913 Niels Bohr demonstrated the possibility of explaining the hydrogen spectrum, the simplest of all spectra, on the basis of the quantum theory; and beyond that of understanding the general Ritz principle through this same theory. Since then the extremely rich development of Bohr's ideas showed wonderful fruitfulness. With the close reciprocal action of theoretical thought and further experimental research our knowledge of spectra increased to a tremendous extent; simultaneously also the meaning of their obvious regularities gradually became clear. To grasp a few of the principal points of this development it will again be necessary to disregard historical sequence.

The so-called "electronic-impact experiments" of Franck and Hertz led to conclusions as fundamentally surprising as the discovery of light quanta. In these experiments the atoms of a gas or vapour were bombarded with electrons, the velocity of which was controllable and known precisely and it was shown that with very slow electrons the collisions are always strictly elastic.

If any macrophysical system capable of internal oscillation is struck a blow from the outside, the system enters into more or less vigorous oscillation. The energy content which the system removes from the impinging body and absorbs as internal energy can assume all possible values from zero to the total energy of the striking body; a very weak blow from an energy-poor body can only transfer a little energy; but there always remains the possibility of a second, weaker collision wherein a still smaller amount of energy is transferred. In all cases it is "infinitely improbable" that no energy at all is transferred; thus, the collision is exactly "elastic",

The Franck-Hertzexperiments show that an atom behaves quite differently in this respect. Electrons of very small velocity can not convey internal energy to the atom at all; an atom never takes up such a small amount of energy. The electron must possess a certain minimum amount of energy if the atom is to take up any from it in its internal system and then it removes all this energy from the electron, never only a part of it. If the kinetic energy of the electron is a little greater than this minimum the atom always withdraws from the hitting electron (if at all) only exactly this minimum amount of energy. If the electron has still greater energy the atom can take from it at collision certain (definite) greater quantities of energy.

In the macrophysical system the energy content has a certain value which can vary continuously; but in the atom the energy content is not capable of continuous change. Instead there exist definite "energy levels" for the atom. The meaning of this phrase will become clear directly the difference between continuously and discontinuously changing energy through comparison of an inclined, smooth path with a staircase. But one must not assume that the energy levels (steps) of an atom follow each other with equal separation. As a matter of fact these separations vary within an atom; and besides, the position of the various energy levels is quite dissimilar in the different elements.

Each element possesses an infinite number of such levels, compressed closer and closer together on top to a certain limit; should the atom take up more energy than corresponds to this upper limit of energy levels it must become "ionized" — an electron is torn out of its electron shell and flies forth with the surplus energy. This ionization of an atom through the impact of a sufficiently high speed electron had already been discovered for a few cases by Lenard; but for us here the energy levels below the "ionization limit" are far more important.

The paradox of this result is obvious: one can not imagine a fact which smacks in the face more brutally all the concepts on which classical physics is based. The principle of continuity is pierced. Since it is established that an atom can never possess energy levels other than those which correspond exactly to those valid for this atom, we must resolve, consequently, that changes of energy also can no longer follow continuously. Therefore the atom is no longer the same as a macrophysical structure, the energy content of which increases and decreases continuously. A change of state through which an atom shifts from one of its possible energy levels to another one is a discontinuous elementary process, a "quantum transition". Nature makes transitions!

Paradoxical as these conclusions are, they agree harmoniously with the equally paradoxical determinations we were forced to make concerning the nature of light. We have now learned that an atom changes its internal energy in discontinuous jumps. We learned earlier that the energy of a light ray — in sharp distinction to the assertions of classical wave theory — is concentrated in individual light corpuscles. Both paradoxes fit together. We arrive at the conception that light production by atoms proceeds in such a way that an atom in a quan tum transition changes its energy discontinuously and emits the energy that has been freed in the form of a light quantum. The absorption of light is the exactly opposite process.

This concept was checked and confirmed in all its inferences by Franck and Hertz (and after them by many other physicists) in an abundance of experimental tests. Through it we can calculate all the spectral lines of an atom from a knowledge of its energy levels; and the converse, since we know that a light quantum of a certain energy (as it results from a quantum transition in an atom corresponding to an energy change of the atom) also possesses a quite definite wave length which can be calculated from the energy according to the above relation. It was also possible to determine directly by experiment that atoms which have been raised to a definite energy level, perhaps by electron impact, emit from the possible spectral lines of the element in question just those for which, according to this concept, the energy level in question is the "initial condition".

Energy transfers between two colliding atoms had also been observed; here simultaneously the one atom jumps to a higher, and the other to a lower level. The possible remaining energy surplus is converted into kinetic energy of the separating atoms. These are only quite fleeting hints; but that the above idea gained from the existence of light producing (or light absorbing) processes is correct has become an irrefutable certainty through modern experimental measurements.

DUALISM; WAVES – CORPUSCLES

We inter rupted our consideration of light phenomena before we succeeded in finding a solution of the contradiction between the wave and corpuscular theories of light which were established reliably through experiments and which both represent inescapable results of experiment. But they mutually contradict each other; and instead of a solution of this contradiction we have 'as yet seen no further than a combination, an arrangement between the two mutually contradictory theories in the form of a relation which permits the proper wave length to be calculated from the energy of a light quantum, or vice versa.

The paradoxes only increase when we realise that this same incomprehensible anomaly also appears in other radiations. We know that cathode rays really consist of a flow of electrons; yet if these cathode rays are passed through crystalline foils, there result interference phenomena analogous to those produced by X-rays.

The discovery of these interference effects was so amazing and unexpected in view of our previous certainty of the corpuscular nature of cathode rays that at first it was disregarded by experimental physicists. The wave properties of cathode rays -like the Maxwell-Hertzian waves — were first predicted by theoreticians. It was de Broglie who arrived at the bold idea that the "dualism" of waves and corpuscles with which we became familiar in light could also be found in cathode rays and other material radiation. De Broglie was able to show that if such an effect is actually present the corresponding wavelengths in the cathode ray must be defined by very simple theoretical laws (for the detection of which the theory of relativity was again extremely important).

Only several years after de Broglie's theoretical disclosures were these paradoxical suppositions checked experimentally. The result was positive; the interference effects predicted by de Broglie were exhibited for cathode rays. Later these experiments were even performed with corpuscular rays — with a stream of atoms; here the proof of the wave properties is still more difficult because the wave lengths are much shorter than for electrons (cathode rays) of the same velocity. According to de Broglie the wave length corresponding to corpuscles at a certain velocity is inversely proportional to their mass. Today we can no longer doubt that this dualism of waves and corpuscles is a quite general physical regularity; each wave radiation which takes place must simultaneously be a corpuscular ray and each corpuscular ray must on the other hand also exhibit wave properties. Practically it is only for light and the lightest corpuscles that it is possible to grasp both sides of the phenomenon with our present experimental methods.

For all other cases the de Broglie mathematical formula yields results for the correspondence of waves and corpuscles. which lie beyond the possibilities of practical observation due to their minuteness. This is just what is to be desired; only in the realms of atomic phenomena is there room for dualistic phenomena; it would be a gross contradiction of experience for a theory to maintain that these dual istic phenomena also occur in the macroscopic, that corpuscles are perhaps evident in radio or sound waves or that conversely interference effects resulted when a machine gun was shot against a garden lattice.

The paradox of this dualism requires no emphasis; here we are faced with phenomena which are completely inconsistent with the ability of our classical physical theories for intellectual reproduction. Let us consider a concrete example.

In a black, light-impervious screen two very small openings are introduced close to each other. Light from a point source passes through these on to a photographic plate set up some distance away. The intensity distribution of the light striking the photographic plate is recorded on it. For sufficiently small openings (and sufficiently small separation between them) the result is not two light spots on the photographic plate as would be expected from exact rectilinear propagation according to geometric shadow construction; the two light rays which come through the two openings interfere with each other.

There was discussion of the supposition that the intensity distribution on the photographic plate would change if the light intensity were made infinitesimally small — so small perhaps, that on the average only one light quantum would be emitted from the source per second. For it is naturally an obvious expedient to attempt to explain the interference effects as a result of a mutual action of different light quanta upon one another.

One would then have to imagine that the light quanta which passed through the one opening reacted mutually with those which went through the other opening, the interference resulting from their interaction. But the inevitable consequence that the interference effect must be destroyed for very small light intensities where the light quanta occur individually and can not combine in any way was not confirmed by experience. Regardless of whether the photographic plate is very strongly illuminated through the screen openings for a very short time or whether it is illuminated for a correspondingly longer time with weak intensity exactly the same diffraction pattern is formed on it.

Thus we can not do otherwise than imagine that interference laws apply to the individual light quantum. Naturally a single light quantum emitted by the source can only be absorbed in a single grain of the photographic plate. Intensity distributions which appear continuous to the rough view can therefore occur only if large numbers of light quanta are absorbed.

Here we must revert to the concept of probability; we must say that one single light quantum emitted by the source possesses a certain probability that it will appear at the exact position of the plate we are viewing; and this probability is given precisely by the light intensity at the point of the plate in question calculated according to classical wave theory. But how does the light quantum reach there from the light source? Previously we emphasized that a self-evident provision of classical Galileo-Newtonian mechanics lies in the conviction that a physical body cannot reach a place from another except by traversing a continuously connected path between these two points.

But how can this be reconciled with the interference considered in our example — from which we know explicitly that interference is not only shown for a light quantum, but likewise appears in principle for material particles like electrons and atoms? The answer we must resolve to accept is that this cannot be reconciled and that we must regard this selfevident provision of classical mechanics as a proposition in plain disagreement with atomic and quantum physics. It is only possible to define the path along which a particle moves continuously insofar as interference phenomena are absent; where interference stands out perceptibly the usefulness of this classical concept ceases fundamentally.

As enigmatical as these facts are and as much as they contradict all our thought and visualization habits, it should be understood that in a certain sense the picture of nature is simplified by this dualism. Formerly we believed that there were both wave and corpuscular radiations in nature; and our classical viewpoint let us consider them as completely and irreconcilably different. Now we see that, in reality, nature recognizes only one kind of radiation which could not be imagined on the basis of classical physics, since on the one hand it exhibits properties which correspond to our classical wave representation but on the other hand corresponds to the classical representation of a corpuscular ray.

We admit that we still have not understood the thing, and make it clear that from these determinations the previously disclosed phenomena become more understandable — "more understandable" in the sense that in any case we recognize connections between them and the paradoxes being discussed now. It has been mentioned that the spatial extent of both the atomic nucleus and the electrons is about a hundred thousand times smaller in diameter than an atom — the hydrogen atom for example.

Now we resume consideration of a problem introduced then — how it is possible that the one electron in the hydrogen atom, despite its small size, can "fill up" the relatively enormous space of this atom. According to our present knowledge the electron must also be imagined as a wave and this problem becomes quite different. Now we must imagine that the charge of the electron, in the sense of the de Broglie wave theory, is somehow "smeared" over the entire volume of the atom, so that this single electron actually forms a "cloud" of electrical charge.

There is also a fact connected with the dualism of waves and corpuscles which in a suitable characteristic manner distinguishes the modern quantum theory corpuscular concept from that of the indestructible atom in Greek philosophy. Electrons (and the same holds for all other corpuscles) possess no "individuality".

Two widely separated electrons may approach each other, meet, separate again and return to their original positions. It can never be determined whether "the same" electron has returned to that position or an "exchange" of the two has taken place — it must immediately appear senseless, considering the previously described criticism of physical statements, to pose this problem at all, to desire a yes or no answer to it. For fundamentally all criteria are lacking which could lead to a choice between the alternatives. We must imagine all electrons as completely equal; and it is not possible to place an "identification mark" on an electron in any way. Thus far we remain in harmony with classical atomic philosophy. But now, through the dualism of waves and corpuscles, a new idea appears. We are no longer to remain certain of the identity of an electron permanently by observing its motion; when the definition of a path becomes impossible or uncertain through interference effects, two electrons which approach each other very closely can "interchange" so that they can no longer be distinguished individually.

THE LIMITS OF CAUSALITY

Consideration of interference phenomena caused a modification of the law of motion of a corpuscular particle traveling free of force. According to Galileo and Newton we said earlier that a body that is not influenced by an external force continues along its path rectilinearly with the velocity it once attained.

If this were the case for light quanta, obviously there would be no interference, no diffraction phenomena; light quanta striking a perforated screen would quite simply be partially kept back by the screen and partially transmitted through the openings, and beyond the openings they would move on rectilinearly — the simple geometrical laws of shadow construction would hold with unlimited exactness, without being broken down by diffraction phenomena. Instead it was necessary to formulate the law of motion for light quanta emitted by the source which reached the photographic plate through the screen openings so that it included the word "probability".

We spoke of the probability that the light quantum in question appears at a certain place on the plate, and formulated the natural law concerning this so that the light intensity calculated according to the wave theory was indicated as an exact measure of this probability. Apart from all other difficulties, this result was in abrupt opposition to classical causality which had attained such clear and convincing form in the Galileo-Newton mechanics.

That we actually do not progress if we adhere to the classical theory of causality is also indicated by various other facts. The radiation which emanates from radium, as the elementary process, is an indication of the disintegration of the radium atom nucleus; emission from the radium nucleus is always in the form of an alpha-particle (helium nucleus) and what remains is the nucleus of another element (radon — radium emanation).

But in a large quantity of radium atoms a simultaneous disintegration of all the radium nuclei does not occur; the law prevails that after 1580 years one half of the original amount of radium has remained unchanged; after 1580 more years half of that half remains, and so on. The physicist finds himself here in the same position as the director of a life insurance company.

Without being able to betray anything to the insured individual about the probable instant of his death, the insurance director can still recognize the statistical law for the average time of death among a large number of insured — and thus can recognize reliably that an insurance business can be established on that basis. Likewise, the physicist knows how many of a thousand million radium atoms, for example, now present will disintegrate within the next year; he does not know whether a single radium atom presented to him will decompose within the next second or will still survive for millions of years.

This could simply be considered as the incomplete state of our knowledge; it is possible to believe that physicists of the future will learn to place a prognosis of its life-expectancy on a single radium atom. But the above stated mathematical law for the "dying" of radium atoms contradicts this idea. A simple mathematical consideration shows that this law is equivalent to the following determination: a single radium atom, submitted to us today, has a definite probability of disintegrating within the next 24 hours. If this radium atom actually has not disintegrated by the first of January of the year 3000, the same probability exists then for its decomposition within the next 24 hours as now.

The problem of radium atoms is entirely different from that of the human being, who, having nearly reached the age of 100 years, must expect his death within the next month with much greater probability than a 20 year old. It is not a peculiarity of living organisms that the probability of death changes in the course of time; it is merely a general result of the theory of causality that we expect the probability for the disintegration of a physical structure due to internal causes to change in the course of time, since it is determined by the previous history of this structure. By contradicting this expectation the radium atoms represent a physical event that is not consistent with our classical conception that each effect can be traced back to a definite cause.

Let us consider still another example that makes it even clearer that this denial of the classical concept of causality is not to be understood as a temporary imperfection of our knowledge, but is inherent in the nature of the thing — again showing how incorrect our previous, classical concepts were. Light waves were previously described as transversely vibrating waves.

With the aid of a "Nicol prism" the physicist can resolve a light ray into two component waves which are "linearly polarized"; that means that in each of these component rays the electric field intensity of the light oscillates only within a definite plane determined by the light ray. In the two component rays these "vibration planes" are perpendicular to each other; one component wave is transmitted by the Nicol prism, the other is reflected. If a linearly polarised light ray impinges on a Nicol prism again, it is resolved in general into two new linearly polarised rays, the plane of vibration of which are mutually perpendicular but are inclined to the plane of vibration of the first ray.

The relation between the intensities of the two component rays after this second resolution depends upon the angle (given by the relative positions of the two Nicol prisms) of this inclination. If we imagine this experiment performed with just a single light quantum we must say that this light quantum has two possibilities — it can be transmitted or reflected through a corresponding rotation of its original plane of vibration; the intensity relation calculated from the wave theory of light must again give the probability of realization of each of the possibilities.

If we wish to maintain the concept that somehow it is causally determined in advance which of the two possibilities will be realized for this one light quantum we shall become hopelessly confused. The angle of inclination of the Nicol prism can be a quite arbitrary one; and besides we can arrange an arbitrarily long row of differently inclined Nicol prisms behind each other; it is impossible to determine the hypothesis of which hidden property of the light quantum predetermines causally how the light quantum will behave in each possible case of this kind without contradicting the irrefutably established probability law for transmission or reflection expressed above.

Thus we must decide to admit that causal predetermination of the behaviour of an individual light quantum near a Nicol prism is not given in nature; nature does not effect the distinction before the occurrence of transmission or reflection.

THE QUANTUM THEORY DESCRIPTION OF NATURE

QUANTUM MECHANICS AND WAVE MECHANICS

We have traversed the realms of physical research as rapid travelers. Since we failed to take along the heavy weapons of mathematics, we have been limited, so to speak, to the role of a civilian war correspondent in the land of physical research who must be content to draw some mood pictures of the events there without seriously pursuing the strategical and technical problems.

We passed by rich departments of knowledge, varied and beautiful in their content. To omit consideration of the thought structures in which modern physical thought triumphs over the apparently hopeless paradoxes and difficulties which were evidenced in quantum physical experiments would require still greater decision. It is not possible to relate and explain here the heavy weight of experimental quantitative proof of the content and mathematical laws of modern quantum theory; a clarification of the philosophical-logical character of this theory by intimation must suffice. But even for that the author, who wants to make it more easily accessible to the reader, must request special attention.

Let us consider briefly the historical development of modern quantum theory. The investigation of atomic spectra (and the energy levels related to them) presented physical research with an abundance of varied problems, in the step by step clarification of which physicists gradually worked out an understanding of the regularities of quantum physics. Niels Bohr, who started this development in 1913 also remained its leader.

Without going into the thousand different problems of this field we shall consider the Bohr "correspondence principle". Although it was mentioned in reference to the energy or relativity principle – and not opposing these in its meaning – the correspondence principle has a quite different character from the energy and relativity principles. These latter are natural laws in finished, perfected form; their content can be expressed clearly and thoroughly in a few words. But the correspondence principle presents peculiar difficulties to the intelligence since its general content can only be described by intimation, in special individual examples. For the correspondence principle is not a finished, clearly definable law of nature, but indicates the direction in which Bohr's conviction about the solution of the quantum puzzle was to be sought, and in which he wished to steer the reflection of the quantum investigators.

It only becomes clear in the history of quantum theory that not only individual great discoveries are decisive for the development of our knowledge but that under certain circumstances the change of our spiritual attitude toward problems can be much more important. Physical thought develops not only from combining the compilations of results of individual observations and their logically-exact treatment; but rather the attainment of decisive new intelligence depends essentially on a creative phantasy, which for its part depends essentially on spiritual hypotheses and on the attitude our intuition assumes toward these things. Bohr's historical contribution lies not only in the individual, pioneer discoveries through which he enriched quantum theory but also in the penetrating force his spirit exerted for the creation of a new spiritual "atmosphere" — wherein the problems were first gradually elucidated as the essential became distinguishable from the unessential and effective control of the solution became possible.

The meaning of the correspondence principle is not purely historical; it is still indispensable today in the sense that it teaches us to view modern quantum theory knowledge with the proper attitude.

The differences between quantum theory and classical physics are so deeply rooted that many physicists were tempted to regard the ideas of classical physics as absolutely useless for the comprehension of atomic physics; they attempted to introduce arbitrarily invented, new, untraditional ideas. But Bohr had — and this is the content of the correspondence principle — nergetically pointed out that in all individual problems a verifiable close similarity exists between classical and quantum theory despite their fundamental difference.

The following is a rough example: classical physics predicts that atoms, which con sist of electrically charged particles, necessarily emit light in the execution of internal motions. This prediction is correct. The elementary processes of this light emission do proceed quite differently from macroscopic antennae; but the fact remains that the fundamental expectation is fulfilled. That is an example of the close relationship which exists between classical and quantum theory despite their incisive difference. There are other examples of this relationship, much finer examples; and through Bohr we gradually learned to see that such kindred relationships can absolutely be uncovered for each concrete individual problem through more exact analysis.

In his emphasis on the affinity of classical and quantum theory there exists, however, a decided prominence of the independence of the quantum theory from classical theory. Since we learn to "understand" various quantum-physical individual problems better through uncovering kindred relationships "in the manner of correspondence", we gradually attain one of the concepts of quantum physics used in classical theory but separated from it by fundamental differences and independent in itself. Such a discovery is a radical challenge to the repeated spasmodic attempts throughout the course of historical development to somehow want to "explain" the characteristic quantum phenomena — to want to reduce them to ideas which conform more closely to the classical.

With the progress of this development it gradually became clear which problems in general should be answered by a logically effected quantum theory. Here the positivist consideration must be repeated very forcefully; the goal of a purely intellectual comprehension of quantum phenomena must be a description of the experimental facts themselves. The experimental facts involved here are collectively of the following sort: a specific atom, characterized by its atomic number, according to experience possesses quite definite energy levels.

The first problem appears; how can the positions of all these energy levels (for any atom, and likewise for molecules) be determined and calculated from general theoretical principles? Furthermore, we find "transition-probabilities" for these atoms. When an atom exists in an energy-rich condition, after some time it will undergo a quantum transition which brings it down to a lower energy level while the energy freed thereby comes off as a light quantum.

The atom has a range of various such possibilities since there are different lower energy levels at its disposal. As in the case of the radium atom described above, it is not possible to predict in an individual case when and where the atom will jump. But experience indicates the existence of quite definite probabilities for the various possible processes. If the atom is irradiated with light there results a definite probability for a quantum transition associated with the absorption of a light quantum. If an electron of a certain (sufficiently high) velocity impinges on an atom, the atom can, as we already know, thus also be induced to a quantum transition. If, furthermore, the atom collides with another atom (of the same or a different kind) an energy change can occur, as was already mentioned, in the form of a simultaneous mutual quantum transition.

Definite probabilities prevail for all these processes; based on experience they are always stated in the same ways through the conditions of the experiment in question. They are summarized under the designation of "transition probabilities". Now the problem (besides the theoretical determination of energy levels) that a complete quantum theory must solve quite comprehensively is simply the theoretical determination of the transition probabilities.

In this sense Heisenberg undertook the creation of a "quantum mechanics". He relied upon the success then already achieved in the detailed execution of Bohr's correspondence principle; in fact, the systematic evaluation of Heisenberg's extensions (Heisenberg,

Born, Dirac, Jordan) yielded a solution of the problem formulated according to the correspondence principle which was complete in principle. Of course the mathematical formulation of this solution was such that it differed extensively from the mathematically precise renditions of our classical physical conceptions. But this mathematical form of the new quantum mechanics was most exactly suited to the problem (as we stated it) of the determination of transition probabilities. Besides, this new quantum mechanics expressed wonderfully the two-sided relation of quantum to classical theory; namely, fundamental difference on the one side and close connection on the other.

These investigations had just reached a preliminary rounding-off point when Schrödinger arrived at the same goal by an entirely different method. Schrödinger started with de Broglie's investigations. After de Broglie had shown how the uniform rectilinear motion of a corpuscular particle was reinterpreted in the wave theory, Schrödinger wondered how these de Broglie considerations developed for motions influenced by external forces. For this reason he investigated the motion of an electron under the influence of the attraction of a positive heavy nucleus; and thus he arrived at the quantitative description of the electron charge cloud in the hydrogen atom.

Schrödinger could now show that with the solution of the mathematical problem he formulated he had simultaneously also achieved the solution for the apparently quite different mathematical problem which was expressed by quantum mechanics in the form of matrix theory. Thus, when a problem has been solved by means of Schrödinger's "wave mechanics" a mathematical conversion yields the "quantum mechanics" solution for this same problem.

This mathematical connection between the two theories, leading to the same result, which in view of the complete difference of the two methods must appear very surprising at first, also supplied the certain basis for the physical-abstract interpretation of Schrödinger waves.

At first it might appear that the discovery of wave mechanics yielded a mitigation of the contrast between quantum and classical theory; for here instead of the unclassical concepts of transition probabilities, etc., we deal with waves — something closer to classical ideas. But what we just learned with respect to light is still true; the classical wave theory is not the final word. Wave mechanics does not in any way signify for atomic physics the removal or mitigation of the fundamental unclassical characteristics of quantum physics. Also, for atoms with more than one electron (thus for all atoms except hydrogen) Schrödinger wave mechanics assumes a very abstract form; in these cases Schrödinger waves are no longer waves in customary threedimensional space but are simply a mathematical construction which mathematicians can "illustrate" to themselves as waves in space of more than three dimensions.

This abstract, multidimensional space can be avoided through another (mathematically equivalent) method of representation, the construction of which was a special hobby of the author's. In this method of representation ("second quantization"), which clings especially closely to the fundamental dualism of waves and corpuscles, the waves dealt with are spread out in ordinary three-dimensional space, but can only be described by means of the ideas of quantum mechanics.

These theories, whose abstract nature will not remain hidden from the reader in even this fleeting explanation, made possible a number of special uses for individual problems of atomic physics. The precise correctness of the new theory was confirmed without exception in a tremendous field of experimental results. No end is in sight for the further elaboration of special problems on the basis of these principles. The fundamental result may be stated as follows: today we can understand all phenomena which occur near and in the electron shells. Thus all the elementary processes that are the basis for spectral or chemical processes are defined as clearly as the motions of the planetary systems have been since Newton. The only partially explored realm at present remains the physics of internal events in atomic nuclei.

For us it is only important to understand the philosophical nuclei of these new ideas. We again owe special thanks to Bohr and Heisenberg for the philosophical-epistemological explanation of the meaning and significance of the new theories. This became possible through the earlier thorough mathematical understanding of the quantum laws in the so-called "statistical transformation theory" (Dirac, Jordan).

OBJECTIVITY

We return once again to classical theory to emphasize still more clearly the characteristic features through which it differs from quantum theory. We can state three basic principles which should be considered as the most important characteristics of classical theory with the catchwords continuity, causality, objectivity.

We spoke amply of continuity; and we learned that just this principle of continuity constitutes a difference between classical and quantum theory. But now we want to make it clear that continuity is not only something familiar to us, but is really an essential, the omission of which is followed with logical necessity by further fundamental deviations from classical ideas.

The assumed continuity of natural processes is essential for our method of executing and evaluating physical measurements. Every measurement we perform is inexact. The physicist considers the "limit of error" for each measurement; for every measured numerical value he records two numbers, the difference between them corresponding to the uncertainty of the measurement.

That it is nevertheless possible to draw certain conclusions from an inexact measurement – conclusions, of course, that are also affected by an uncertainty but in any case are rich in content and very definite – is only because the principle of continuity is actually fulfilled in the macrophysical world; trivial changes in the cause are followed by trivial changes in the effect. Consequently an inexact knowledge of causes is still sufficient for an inexact prediction of the effects, although the absolutely certain prediction of the effects becomes possible only with a mathematically more precise (practically unattainable) measurement of the causes.

Uncertainty in the process of measurement deserves a still more detailed investigation. The essential point for us here will become clear if we visualize, for example, the measurement of the temperature of a macrophysical body. We bring the body in contact with a thermometer; the thermometer assumes the same temperature as the body in question; and we read off the value. But to be exact, it is necessary to consider that since the body under investigation imparts some of its heat to the thermometer the body itself is influenced and its original condition is altered somewhat.

To avoid resultant false measurements the thermometer selected must be much smaller than the body being investigated. (Of course it is possible, if the thermometer used was not sufficiently small, to compute or estimate the temperature change which took place and correct the result for it; but that is a problem in technical method which has no connection with the nature of the thing.) It is analogous for every physical measurement: I must always select such fine instruments for a measurement that in the measuring process the body under investigation itself will not be influenced appreciably by too rough a measuring instrument (such influence would falsify the measured result). Fundamentally a reaction of the measuring instrument on the object being investigated is inherent in every physical measurement; yet in the investigation of macrophysical objects this reaction can be made sufficiently small by the selection of quite fine instruments. [1]

This idealization of the process of measuring in classical physics depends essentially on the assumption of continuity of all natural processes in the sense that our fundamental considerations are established as though precise measurements were possible for us. It is a "permissible idealization" to base our considerations on this concept of absolute observational accuracy and to interpret the actual inaccuracy of each measurement as an only practically and not principally significant secondary disturbance. This idea cannot lead us to errors, because by classical concepts we can approach ideal measurement unlimitedly, although we can never attain it.

We here consider something which we shall indicate with the word objectivity and which is so characteristic of our general classical-physical thinking that we usually disregard it. We are accustomed to regarding physical observation results as clearly understood according to their importance if we have explained them as effects of an objective physical process or condition.

This formulation hides within it a higher type of positivist modesty; a deeper "explanation" of natural processes is relinquished herein for everything. But what is left after this renunciation of all classical-physical theories as a basis for their methods of representation is just this idea in objective events. Perhaps we do not describe planetary motions by specifying when and where or through what telescope the various planets were observed; but we do describe planetary motions as a spatial-temporal process taking place independently of human observation. Naturally one can pursue the familiar philosophical considerations which emphasize that without subjects doing the observing there can be no talk of objective matters. But these thought processes are not acceptable to the physicist. To him the motion of the planet Neptune in its orbit is an objective event which was already in process before anyone had seen this planet in a telescope and which continues uninfluenced, independent of whether or when or how often it is observed or photographed. Exactly the same self-evident supposition of objective events is the basis for Maxwellian electrodynamics; we interpret electrical measurement data as indication of an objective physical event in the electromagnetic field which is present in space.

Positivist criticism must remind us that this objectivity of physical events is not a purely logical self-evident fact. For positivism teaches us to view true physical reality only in the totality of experimental results. It is very remarkable and astonishing that in the domain of validity of macrophysics we are in a position to so formulate our summarizing description of experimental results that they are no longer referred to directly but are, so to speak, only minor appendages of the picture we have traced out — a picture which maintains the existence of objective events which occur independently of how and where the observations necessary for their detection are taken.

After what has been said it is obvious that the idea of ideal measurements (made possible by the continuity of macrophysical processes) which do not disturb the observed event in the least is in dispensable for this construction of an objective physical world. Another assumption indispensable for this objectivity is that of complete causality in the macrophysical world.

For if we had had to reckon with the occurrences of effects not determined strictly causally we could never have been certain in our observation processes whether the source of an effect we saw was to be sought in the object we were observing or just in a cause-free reaction of our observation instrument.

Thus objectivity would be disturbed if complete causality did not exist; conversely, the representation of objective physical events is a necessary assumption for the carrying through of the idea of strict causality. That complete causality is an indispensable assumption for the possibility of effecting the representation of objective physical events was already clearly recognized by Kant. But there is no justification for concluding from this insight that the complete validity of the principle of causality throughout nature is guaranteed from the outset independently of experimental experience. All that must be established is this — objectivity also is weakened with the renunciation of complete causality.

This is actually the case in atomic physics; we have seen that the principle of continuity can not be given up without objectivity ceasing. And we have learned that continuity ends in atomic and quantum physics. We can consider the appearance of discontinuities in elementary physical processes as the fundamental proposition of modern quantum physics; the atomistic structure of matter may be in terpreted as one facet of this elementary physical discontinuity which also appears in many other forms.

The above examples have already made it clear that in quantum physics a complete causality, of the type we are accustomed to, no longer exists. But we still want to refer particularly to the difference that exists between the quantum theory use of the probability concept and the Boltzmann evaluation of it (for the kinetic explanation of heat processes). In Boltzmann's considerations statistics was a secondary thing; at that time there was no reason to doubt that the motion of each individual atom could fundamentally be calculated precisely in advance.

The pursuit of such fine processes was voluntarily relinquished and statistical considerations were used as an expression of an incomplete (but sufficient for the result desired) observation of the events. Whereas in quantum theory the primary natural laws themselves take the form of probability expressions in this case the statistical concepts are not an expression of the incompleteness of our insight into events, but rather an expression of an indefiniteness existing in nature itself. Nature herself did not determine individual atomic processes in advance; from case to case she executes unpredictable decisions which show fixed regularities only in the statistical average. But these unpredictable decisions of nature are always connected with elementary quantum physics discontinuities. Indeed it is the independent cases of quantum transitions that are not predetermined.

Our earlier example of the interference of light for the screen with two openings already demonstrated that quantum physics must also relinquish the idea of objective events.

With characteristic positivist modesty we limited our problem to the problem of how great the probability is that the light quantum in question will be absorbed in a certain point of the photographic plate. Thus two quantum transitions — and that is typical of the whole modern quantum theory - are placed in a static relation with each other. We observe the quantum transformation of the light emission from the point source; then we observe the quantum transformation of the absorption of the light quantum in a certain grain of the plate; and we can theoretically calculate in advance the probability of the occurrence of the second elementary act after the incidence of the first. But we cannot extend the picture of an objective event between both processes in the form of the specification of a continuous path which the light quantum must traverse from the first position to the second.

Formerly one might have been inclined to suspect that the principle of the objectivity of physical events was also inseparably connected with the possibility of the quantitative, mathematical comprehension of natural regularities. But we see that that is plainly incorrect; everything we try to explain here in words can be expressed in as mathematically clear a form as Galileo-Newton mechanics. We must replace the precalculation of future events on the basis of complete causality possible in the macrophysical domain by a purely proba bility prediction. The probabilities of quantum physics processes are themselves determined exactly quantitatively; in the last analysis they are subject to precise mathematical laws of great simplicity and of the most comprehensive validity.

Despite that, the Kantian interpretation, that complete causality (as well as objectivity and continuity) of natural processes is an indispensable provision of every physical thought in general, is still correct in the following sense: quantum physics experiments are also always performed with macrophysical apparatus. We need macrophysical (thus functioning according to strict causality) apparatus to be able to make any correct observations at all and to be able to determine regularities in the atomic world.

Classical physics remains the indispensable support from which an advance into the .world of quanta and atoms becomes possible. This can not be altered by the fact that macrophysical laws can naturally be interpreted as results of quantum physics elementary laws.

The laws which govern the motions of macrophysical bodies must naturally result from the laws to which their individual atoms are subject. There is no difficulty involved in the necessity of interpreting the strict causality of macrophysical events as a result of the purely statistical laws for the elementary processes. For that is the essence of conformity to statistical laws — despite the incalculability of the separate event, an exact, predictable result occurs in the total effect of a large number of individual processes. The fact that we thus interpret the laws of atomic physics as the really true natural laws

from which macrophysical laws are derived as results does not permit us to overlook the other fact, that the elementary laws of atomic physics include a tangible content only when they are attached to the frame of macrophysics by concrete application. Herein lies in the final analysis the root of the not only historical, but contemporary significance of the Bohr correspondence principle. For this teaches us to understand the meaning and content of quantum physics laws from their relation to macrophysics.

COMPLEMENTARITY

Experimental evidence has shown us in a most comprehensive way and with a variety exceeding all expectations the atomistic structure of all physical substrata; not only matter, but also light (despite its wave nature which it exhibits "on the other side") has a corpuscular make-up.

This evidence drives us to remarkable conclusions. How can one observe and investigate individual atoms? We saw that modern experimental technique permits the very satisfactory performance of experiments immediately involved with individual atoms (and therewith to exactly prove the reality of these atoms, without a doubt). Naturally experimental manipulation of individual atoms remains much more difficult and far different fromthe investigation and measuring of macrophysical, visibly large bodies which are made up of innumerable atoms.

For macrophysical bodies we have scales and other mechanical, optical or electrical measuring instruments at our disposal; but there is a decisive difference for measurements on atoms. We know that every physical substratum, therefore every physical measuring instrument, is composed of atoms — be they material atoms or electrons or light quanta. This destroys all possibilities of using convenient measuring instruments for this research as is done for macrophysical objects or events.

We have already considered how fundamental for our classical physical ideas and methods of perception is the fact that in macrophysical investigations the back-coupling, the influencing of the object by the observation process, can be made negligibly small through the use of sufficiently fine measuring instruments. But if we consider that all measuring instruments themselves consist of atoms (thus can never be made finer and smaller than single atoms) we see that this method of eliminating the reaction of the measuring instrument is barred when the object to be investigated is itself an individual atom (or a structure containing just a few atoms).

There is no longer any possibility of investigating and observing with instruments that are finer than the object in question. Nor is it possible to control the influencing of the object by the measuring instrument and to eliminate it from the result by a corresponding conversion. Thus it must be considered as part of the bargain that fundamentally measurments on atomic objects are always "falsified" in the sense that according to natural law the object experiences a variable interference from the execution of the observation process.

Similarly when we try to observe and psychologically control our own thought processes we are successful up to a certain degree; but the functioning of the observation itself again influences the observed "object" — our own thought process — and with this sets the limits for the possibility of observation. (One can not, e.g, watch through psychological self-observation how one falls asleep, because just that attention of observing prevents one from falling asleep. Also, e.g., the origin of voluntary decisions is disturbed and changed by internal-controlling self-observation.) Similarities are also found in atomic physics — the observed object is influenced by the observation process itself. With Niels Bohr we can say, the separation between observed object and observing subject begins to vanish here.

It is not at all proper to call this influencing of the object by the process of observation a "falsification". For we are not dealing with a disturbing influence which is in any way limited for the present by the current deficiency of our observation technique; these barriers to the possibility of an ideal observation which does not influence the object itself are limited — through the atomistic structure of physical foundations — by natural law. Therefore, we must ascribe to the atomic objects themselves a certain character of "indeterminateness", of "indefiniteness" of their physical behaviour which makes the construction of an objective picture of atomic physical events impossible.

This need not imply that every possibility of exact measurement on atomic objects is doomed to failure. It is possible throughout to make each physical property of an atom the object of precise measurements. If a certain property of an atom is observed exactly there result from the observation process (due to the reaction of the measuring instrument to the object) strong uncontrollable and undefined changes (strong "uncertainties") with regard to other properties of this atom. By optionally transferring the interference which is necessarily associated with observation of the atom to different properties of the object it is possible to make accurate observations on the properties of the atom undisturbed by the particular reaction. With this "complementarity", as Bohr named it, the new theory can eliminate those apparently hopeless contradictions we first encountered in the dualism of waves and corpuscles and which we met at every step in quantum physics.

This idea of complementarity must be viewed as the most significant result for philosophy that crystallized out of modern physics.

It presents an absolutely new scientific way of thinking which is fundamentally different from classical scientific thinking in terms of objectivized representations. After the intellectual penetration and comprehension of atomic physical phenomena completely inaccessible to the previous method of representation became possible through it, it appears justified to believe that it may further become of epoch-making importance in other realms of natural science.

The enlightening force of this idea in solving an apparently insoluble puzzle and contradiction is clearly demonstrated in the famous problem of the dual istic nature of light. The properties connected with the wave nature of light on the one hand and those connected with its corpuscular nature on the other are "complementary" to each other in the sense that they can never appear in one and the same experiment at the same time (and thus come into actual direct opposition).

Experiments which let the wave side of light emerge clearly force (through the action that is connected with every experiment) the corpuscular nature of light back into the indeterminate and unobservable; other experiments, which force the corpuscular side of light into prominence, leave undefined and indiscernible all the properties which usually betray to us the wave nature of light. With this wonderful device of complementarity nature combines in one and the same physical object properties and regularities that contradict each other so that they could never exist directly at the same time.

Let us pursue this in somewhat more detail in the already repeatedly discussed example of the interference of light rays transmitted through two screen openings. If we want to let the interference of these two light rays take place to the wave nature of light we must, as already indicated, relinquish the desire to determine simultaneously through which of the two openings a definite light quantum was transmitted. We can also consider this event from the "complementary" opposite side.

We can undertake to desire to observe through which opening the light quantum was transmitted. But then we must conversely relinquish any desire to obtain an interference of the two rays. For with the latter intention the only procedure we can follow is to roughly close up one of the screen openings in order to be certain that a transmitted light quantum could actually only have passed through the other opening. There is no experimental possibility of assuring ourselves in any way through which of the openings the light quantum was transmitted without simultaneously so altering the conditions of the experiment that the interference of the two rays is hindered.

As another example let us consider the charge cloud of the electron in the hydrogen atom. If we fix the position of a definite point with a microscope we must assume an inaccuracy which is at least of the order of magnitude of the wave lengths of visible light — thus much larger than an atom. But nothing hinders our imagining that we have a microscope that is not dependent on visible light, but on X-rays or even rays of much shorter wave length (gamma rays). With this microscope we could accomplish measurements of position on an electron with such great accuracy that positions inside of the charge cloud of the hydrogen atom would also be distinguished.

Now let us place a hydrogen atom under this gamma-microscope and "examine" the inner structure of this atom so that we determine the position of the electron. In the description of this imaginary experiment we must not forget one thing; namely, that the energy of the "gammalight" necessary to illuminate and render the electron visible is concentrated in individual light quanta which possess very high energy because of the minuteness of their wave lengths.

The procedure of the "examination" of the electron in the hydrogen atom is as follows: a single energyrich light quantum meets the electron, and reflected by it approaches us through the microscope, showing us the exact position of the electron. But in this process the electron experiences a terrific effect; it is thoroughly dislodged from its former condition by the interaction with the energy-rich light quantum and in most cases we can expect that the electron is completely torn away from the hydrogen nucleus; the atom, thus, is ionised.

We can now explain in a concrete way the meaning of the "charge cloud" around the atomic nucleus calculated according to wave mechanics. We repeat the above described experiment innumerable times; and, it must be emphasized, each time we use a hydrogen atom which is in its lowest energy state (normal state). If, instead, we took atoms in a fixed, higher energy level we would have to reckon with a different structure of the charge cloud; for each of the different energy levels, according to quantum theory, there exists a certain specific structure of the charge cloud.

By repeatedly measuring the position of the electron in a hydrogen atom in its normal state we find the electron in different positions from case to case, just as in our interference experiment we found the individual light quanta in different places on the photographic plate, statistically distributed according to the light intensity calculated by the wave theory. The statistical distribution of the individually measured electron positions is given by the charge cloud calculated according to wave mechanics. It was only with this determination that the concept of this charge cloud attained a clear meaning which is defined by concrete, demonstrable experiments.

At the same time we see in this experiment how the changing actions in quantum physical observation are bound up with the observed facts. Before the act of observation in question the hydrogen atom possesses a definite energy, but in general in these circumstances a definite position of the electron does not exist – the position of the electron is undefined, or only indefinitely defined in the structure of the statistical charge cloud.

It is only through the process of observing its location in the gamma-ray microscope that we force the electron to assume a definite position. Notice, we do not prescribe in which position it should emerge; but we do force it into some definite position and thereby force it to a new crisis; now the electron assumes a definite position but simultaneously an indefinite energy exchange has taken place between the gamma light quantum and the electron and the original condition of a defined energy of the atom has been disturbed.

A quite analogous process is represented by the previously described impinging of a linearly polarized light quantum against a Nicol prism. In this case we can also say that the execution of an act of observation has forced the light quantum to assume a clearly defined situation, when previously this was indefinite. The light quantum must decide whether to be transmitted by the Nicol prism which has been placed inclined to the original plane of vibration of the light quantum or to be reflected by it. That is a decision of the same kind as that of assuming a definite position forced upon the electron in the above experiment.

It is obvious that there is no more place among these ideas for a complete causality, clearly predetermining each occurrence. If in a macrophysical structure, the planetary system, for example, we want to pre-calculate future movements exactly we must know two things. First, we must know that the Newtonian law is valid (and not any other one), and we must know the magnitudes of the different planets which are determinative for Newtonian gravitational attraction and for the relations of force and acceleration.

Secondly, for any one point in time we must know the positions and velocities the different planets possess at exactly this point in time. Thereby the general course of motion is mathematically precisely determined for all later (besides, also earlier) times. In an electron, however, we are not able to simultaneously determine its position and velocity at a definite point in time.

Since position and velocity are complementary, the position measurement in the gamma-microscope makes the velocity of the electron unobservable — and the converse also holds. After knowing that the physical properties of an atom are partially complementary to one another, that therefore it is impossible to observe the atom simultaneously "from all sides" as it were (as is possible in macrophysical bodies), it must be regarded as quite natural that precalculations of the future conduct of atoms, electrons and light quanta are not possible analogously as in the planets. Let us emphasize once again: this impossibility not only depends on a practical technical incompleteness of our instruments, but depends on nature itself. It is a positive result of the natural laws which in quantum or wave mechanics have attained a formulation which is mathematically exact and is verified by innumerable experiments.

The indefiniteness inherent in the physical condition of all atomic objects stipulates a corresponding indefiniteness in the process of the action; pre-calculation according to exact causal laws is lacking here. The inability of the physicist to predict for an individual case which of the various possibilities will be realized in the quantum transition which is the basis for an observation process is not due to human imperfection of knowledge; nature herself has reserved until the last the decision for each individual case.

Finally, let us make it clear that our repeated use of the word "indefiniteness" in relation to atomic physical events actually expresses nothing but the impossibility of using familiar classical concepts in the place in question.The impossibility of describing the relations in objectivisedprocess pictures lends as many difficulties to the verbal expression as does the problem of a clear representation.

Chapter 11

Geometry and the Measurement of Length

SPACE AND TIME

From birth to death we live our lives in being and becoming. It is ancient wisdom concerning the life of man, 'so soon passeth it away'. The becoming, the passing, are in time. We pass our lives on the surface of the earth—in the great river valleys, maybe—in the far distances there are the everlasting hills, above the everchanging clouds, and the stars at night in their fixed constellations.

These intimations arise from the immediate deliverances of our senses; our intuitive appreciation is of the flux of time—and of a vast arena of space determined by the stable earth and the stars. In seeking to build up the science of physics on these foundations, we consider separately the two notions which they involve, the notion of time and the notion of space.

PRACTICAL GEOMETRY

Many thousand years ago in the great river valleys of Egypt and Mesopotamia the kings' surveyors mapped their masters' fields, the royal architects planned the temples and tombs and palaces. Practical geometry had already developed into a serviceable body of rules: in relation to the comings and goings of daily life the structure of local space was understood. There was a secure tradition and great uniformity of practice. Evidence from the buildings which remain, or have been excavated, shows clearly that over a score of centuries the master builders of Egypt were employing a unit of length—the cubit—which did not vary by much more than I part in 200 from the mean.

This bespeaks great attention to the preservation of a linear standard. We know from the records that the royal standard of volume—the apet—was for a long time preserved in material form in the dromus of Anubis at Memphis. Somewhat later, as regulating all measurements on the rise and fall of the waters of the Nile, a standard cubit was kept in the capital in priestly custody in the Sepulchre of the Sacred Bulls, and on the road from Memphis to Faium there was marked by official decree the schoenus, an itinerary standard of 12,000 cubits. Nearly a thousand miles away, in Mesopotamia, during much the same period, essentially the same standard of length, the cubit, was maintained, with almost equal exactitude.

ABSTRACT GEOMETRY

Such were the achievements of a practical civilisation three thousand years before the beginning of the Christian era. As that civilisation was passing into slow decline there was developing around the shores of the Aegean sea another civilisation of a very different character. Individual Greeks visited Egypt over many hundred years as travellers, or in pursuit of trade; in 332 BC under Alexander the Great they arrived as conquerors.

Among the early travellers was Thales of Miletus (*c.* 624-547 BC), and a little later Pythagoras (*c.* 572-497 BC). In these two the genius for abstract thought which was to be typical of their fellow-countrymen for another five hundred years was already fully developed.

To the Egyptians (it has been said) the geometrical point was realised only as a mark or position, the straight line only as a stretched string or the tracing of a pole, a surface merely as an area: to Thales it seems these entities took on the much more exciting nature of elements in an artifact of the mind, a formal logic-bound system, a new abstract language in which the perfection of actual space should be exhibited unsullied by the inconsistencies which derive from imperfect measurement. We speak of that formal system, that new language, as abstract geometry.

From Thales, through Pythagoras and the great Greek mathematicians, Eudoxus perhaps the greatest of them all, the subject received successive additions and refinements until it was systematised by Euclid, the founder of the mathematical school of Alexandria—that new city which the Macedonian conqueror had caused to be built on the delta of the Nile. Euclidean space was accepted as satisfying in every particular the intuitive notion and the accumulated experience of man at large in the world; it was accepted in consequence as providing the spatial framework of mathematical physics.

DEFINITIONS AND AXIOMS

The theorems of Euclidean geometry, by which the detailed properties of figures in space are derived, are based on definitions, axioms and postulates. The definitions draw attention as precisely as possible to the typical entities out of which the figures may be regarded as built up—points, lines, surfaces, geometrical solids, angles. The axioms record what is commonly agreed regarding these entities.

Any set of axioms should consist of individual statements which are mutually consistent, each one being independent of all the others: the axioms of Euclidean geometry have the further character that what they accept as assumption in respect of Euclidean space is commonly believed to be true of the space of experience.

Thus, whilst we should distinguish at the outset between the solid bodies ('rigid' bodies) of physics and the geometrical solids of Euclid, there is, nevertheless, the underlying belief that the mere properties of extension of physical bodies are the properties attributed to, or formally derivable for, the solids of abstract Euclidean geometry.

The postulates of Euclid are not uniquely distinguishable from the axioms. They are of the same nature, in that they are assertions which are unprovable within the logical system to which they belong. They differ from the axioms only in that they are the less obvious of the assumptions which must be made in order to remove all the ambiguities inherent in the situation.

If it were not for the requirement that Euclidean space should be the exact formal counterpart of the space of experience, there might be a basis of distinction; when this requirement obtains, however, definitions and agreed assumptions regarding the entities defined should be sufficient basis for the abstract geometry. We therefore make no further distinction here between axioms and postulates.

STRAIGHT LINES AND PLANE SURFACES

Euclid adopted the synthetic method, building up by definition from the point, the ultimate abstraction—position in space, devoid of extension—through the line and the surface to the geometrical solid, a finite portion of space, completely cut off from the rest of space—by a continuous surface, or by surfaces intersecting in lines—having the fullness of possible extension, in length and breadth and thickness.

The particular notions of the straight line and the plane surface are more difficult to formulate in the abstract than they are in the practical geometry of experience, just because they are particularisations of the general notions 'line' and 'surface'.

To the Egyptian surveyors the stretched string was an obvious real adjunct to their art—and a plane surface was to be found wherever a large expanse of still water was unruffled by the wind. The paving-stones for the temple courts were worked plane, and laid in position so that the courtyards themselves should have the same character.

The intuitive approach to the notions 'planar' and 'straight' was direct. Let us here, in the general Euclidean tradition, attempt the abstract approach: 'a straight line is uniquely determined by any two points in it'; 'a plane surface is such that the straight line through any two points in the surface lies wholly in the surface', or, alternatively, 'a plane surface is uniquely determined by any three points in it, provided that the three points do not lie in the same straight line'.

MEASUREMENT OF LENGTH IN ABSTRACT GEOMETRY

In relation to physical measurement the most important axiom of Euclidean geometry is one which was implicitly assumed by Euclid, but not explicitly stated: 'figures may be freely moved in space without change of shape or size'. This is the counterpart of our intuitive belief in the permanence of the rigid bodies of the physical world. We move our measuring rods from place to place and we assume that they do not change in any way in the process. In particular we may move any segment of a straight line along that line, say from left to right, setting it down successively in such circumstances that the left-hand extremity of the segment coincides with the position previously occupied by its right-hand extremity. In such a way we are able to produce, in the abstract, a rectilinear scale, equally divided, providing multiples of any arbitrary length as unit length; we are able, in fact, to construct a segment of a straight line the length of which is any integral multiple of a chosen unit. We may then move this rectilinear scale as we will, and, by bringing it into coincidence, along its length, with any other straight line, we may estimate the length of a segment of that line, with an uncertainty less than one scale unit, by the simple process of counting.

To estimate length more closely than this we obviously require a method of subdividing the segment chosen as unit in our previous consideration into a whole number of sub-multiples. Such a method is available on the basis of Euclid's theorem that a line drawn parallel to one side of a triangle cuts the other two sides proportionally. So far with Euclid, but abstract geometry is not subject to the limitations of magnitude which determine the practicability or otherwise of actual manipulations of bodies in the physical world.

Thus we can extend, in thought, the process of subdivision of our linear scale to whatever degree of fineness we can imagine, and so provide ourselves with the means of estimating the length of a straight-line segment (or the distance between two points) to within any specified residual uncertainty. Let us consider the operations involved if such an estimate is to be expressed in the decimal notation of Stevinus (1548-1620) and Henry Briggs (15561630). The rectilinear scale, uniformly divided to give multiples of the unit, is brought into coincidence with the straight-line segment of unknown length. One of the points by which the scale is divided is made to coincide with one end of the segment, and it is found, by counting, that the other end of the segment lies in the (A+ 1)th unit interval along the scale, *A* being the appropriate integer.

The (A + 1)th unit interval is then divided into 10 equal sub-intervals by the method just described. When this is done, the end of the segment is found to lie in the (a + 1)th sub-interval, let us say, a being another integer. The (a + 1)th sub-interval is therefore divided into 10 equal smaller intervals and the position of the end of the segment in these is determined, as in the (b + 1)th smaller interval, say, with b integral. The process of successive subdivision, location and counting is continued until the residual uncertainty is less than the originally specified small quantity. The length of the unknown interval is in fact $A{\cdot}abc$. . . units of length, to as many places of decimals as required.

If the segment of unknown length had been the diagonal of a square of which the side had been taken as unit length, we know from Pythagoras's theorem that the process of 'measurement' would have yielded the result: length of diagonal = 1·41421356 sides. And it is clear that in imagination, that is with full validity in the abstract, the geometrical procedure in this case is no more limited than is the arithmetical procedure of extracting the square root of 2. And so in general: in the abstract the problem of determining the distance between two points has been solved, once the unit of length has been chosen, at least as long as we are satisfied with an end-result which is inexact in the sense that the ratio of incommensurables (as an arbitrarily chosen unit length and any other length must generally be) cannot be exactly expressed in the decimal notation by digits in a finite array. We have now to consider the problem in practical geometry and the limitations therein encountered.

Chapter 12

Special Relativity

GALILEAN RELATIVITY AGAIN

At this point in the course, we finally enter the twentieth century— Albert Einstein wrote his first paper on relativity in 1905. To put his work in context, let us first review just what is meant by "relativity" in physics. The first example, mentioned in a previous lecture, is what is called "Galilean relativity" and is nothing but Galileo's perception that by observing the motion of objects, alive or dead, in a closed room there is no way to tell if the room is at rest or is in fact in a boat moving at a steady speed in a fixed direction. Everything looks the same in a room in steady motion as it does in a room at rest.

After Newton formulated his Laws of Motion, describing how bodies move in response to forces and so on, physicists reformulated Galileo's observation in a slightly more technical, but equivalent, way: they said *the laws of physics are the same in a uniformly moving room as they are in a room at rest*. In other words, the same force produces the same acceleration, and an object experiencing no force moves at a steady speed in a straight line in either case.

Of course, talking in these terms implies that we have clocks and rulers available so that we can actually time the motion of a body over a measured distance, so the physicist envisions the room in question to have calibrations along all the walls, so the position of anything can be measured, and a good clock to time motion. Such a suitably equipped room is called a "*frame of reference*"—the calibrations on the walls are seen as a frame which you can use to specify the precise position of an object at a given time. Anyway, the bottom line is that no amount of measuring of motions of objects in the "frame of reference" will tell you whether this is a frame at rest or one moving at a steady velocity.

What exactly do we mean by a frame "at rest" anyway? This seems obvious from our perspective as creatures who live on the surface of the earth—we mean, of course, at rest relative to fixed objects on the earth's surface. Actually, the earth's rotation means this isn't quite a fixed frame, and also the earth is moving in orbit at 18 miles per second. From an astronaut's point of view, then, a frame fixed relative to the sun might seem more reasonable. But why stop there? We believe the laws of physics are good throughout the universe. Let us consider somewhere in space far from the sun, even far from our galaxy. We would see galaxies in all directions, all moving in different ways.

Suppose we now set up a frame of reference and check that Newton's laws still work. In particular, we check that the First Law holds—that a body experiencing no force moves at a steady speed in a straight line. *This First law is often referred to as The Principle of Inertia, and a frame in which it holds is called an Inertial Frame.* Then we set up another frame of reference, moving at a steady velocity relative to the first one, and find that Newton's laws are o.k. in this frame too. The point to notice here is that it is not at all obvious which—if either—of these frames is "at rest".

We *can*, however, assert that they are both *inertial* frames, after we've checked that in both of them, a body with no forces acting on it moves at a steady speed in a straight line (the speed could be zero). In this situation, Michelson would have said that a frame "at rest" is one at rest relative to the aether. However, his own experiment found motion through the aether to be undetectable, so how would we ever know we were in the right frame?

As we mentioned in the last lecture, in the middle of the nineteenth century there was a substantial advance in the understanding of electric and magnetic fields. (In fact, this advance is in large part responsible for the improvement in living standards since that time.)

The new understanding was summarized in a set of equations called Maxwell's equations describing how electric and magnetic fields interact and give rise to each other, just as, two centuries earlier, the new understanding of dynamics was summarized in the set of equations called Newton's laws. The important thing about Maxwell's equations for our present purposes is that they predicted waves made up of electric and magnetic fields that moved at 186,300 miles per second, and it was immediately realized that this was no coincidence—light waves must be nothing but waving electric and magnetic fields. (This is now fully established to be the case.)

It is worth emphasizing that Maxwell's work predicted the speed of light from the results of experiments that were not thought at the time they were done to have anything to do with light - experiments on, for example, the strength of electric field produced by waving a magnet. Maxwell was able to deduce a speed for waves like this using methods analogous to those by which earlier scientists had figured out the speed of sound from a knowledge of the density and the springiness of air.

GENERALIZING GALILEAN RELATIVITY TO INCLUDE LIGHT: SPECIAL RELATIVITY

We now come to Einstein's major insight: the Theory of Special Relativity. It is deceptively simple. Einstein first dusted off Galileo's discussion of experiments below decks on a uniformly moving ship, and restated it as : *Einstein then simply brought this up to date,* by pointing out that the Laws of Physics must now include Maxwell's equations describing electric and magnetic fields as well as Newton's laws describing motion of masses under gravity and other forces.

Demanding that Maxwell's equations be satisfied in all inertial frames has one major consequence as far as we are concerned. As we stated above, Maxwell's equations give the speed of light to be 186,300 miles per second. Therefore, *demanding that the laws of physics are the same in all inertial frames implies that the speed of any light wave, measured in any inertial frame, must be 186,300 miles per second.*

This then is the entire content of the Theory of Special Relativity: the Laws of Physics are the same in any inertial frame, and, in particular, any measurement of the speed of light in any inertial frame will always give 186,300 miles per second.

Just as Galileo had asserted that observing gnats, fish and dripping bottles, throwing things and generally jumping around would not help you to find out if you were in a room at rest or moving at a steady velocity, Einstein added that no kind of observation at all, *even measuring the speed of light across your room* to any accuracy you like, would help find out if your room was "really at rest". This implies, of course, that the concept of being "at rest" is meaningless. If Einstein is right, there is *no* natural rest-frame in the universe. Naturally, there can be no "aether", no thin transparent jelly filling space and vibrating with light waves, because if there were, *it* would provide the natural rest frame, and affect the speed of light as measured in other moving inertial frames as discussed above.

So we see the Michelson-Morley experiment was doomed from the start. There never was an aether wind. The light was not slowed down by going "upstream"—light *always* travels at the same speed, which we shall now call c,

c = 186,300 miles per second,

to save writing it out every time. *This now answers the question of what the speed of light, c, is relative to.* We already found that it is not like sound, relative to some underlying medium. It is also not like bullets, relative to the source of the light (the discredited emitter theory). *Light travels at c relative to the observer,* since if the observer sets up an inertial frame (clocks, rulers, etc.) to measure the speed of light he will find it to be c. (We always assume our observers are very competent experimentalists!)

TRUTH AND CONSEQUENCES

The Truth we are referring to here is the seemingly innocuous and plausible sounding statement that all inertial frames are as good as each other - the laws of physics are the same in all of them — and so the *speed of light* is the same in all of them. This Special Theory of Relativity has some surprising consequences, which reveal themselves most dramatically when things are moving at relative speeds comparable to the speed of light. Einstein liked to explain his theory using what he called "thought experiments" involving trains and other kinds of transportation moving at these speeds (technically unachievable so far!), and we shall follow his general approach.

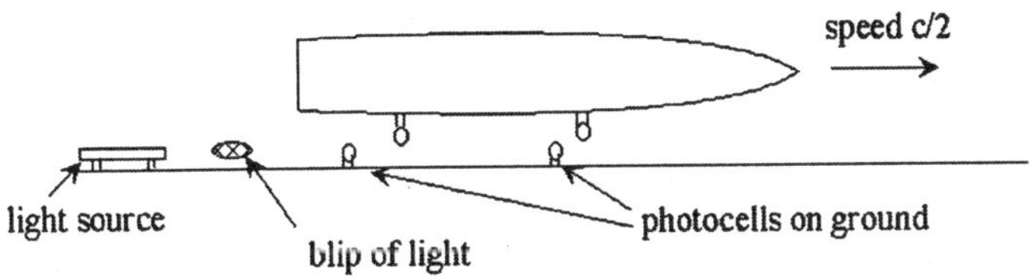

To begin with, let us consider a simple measurement of the speed of light carried out at the same time in two inertial frames moving at half the speed of light relative to each other. The setup is as follows: on a flat piece of ground, we have a flashlight which emits a blip of light, like a strobe. The photocells are placed a known distance apart in the path of the blip of light, they are somehow wired into a clock so that the time taken by the blip of light to travel from the first photocell to the second, in other words can be measured. From this time and the known distance between them, we can easily find the speed of the blip of light.

Meanwhile, there is another observer, passing overhead in a spaceship traveling at half the speed of light. She is also equipped with a couple of photocells, placed a known distance apart on the bottom of her spaceship and she is able to measure the speed of the same blip of light, relative to her frame of reference (the spaceship). The observer on the spaceship will measure the blip of light to be traveling at c relative to the spaceship, the observer on the ground will measure the same blip to be traveling at c relative to the ground. That is the unavoidable consequence of the Theory of Relativity.

The Laws of Physics are the same in any Inertial Frame of Reference. (Such frames move at steady velocities with respect to each other.) These Laws include not only Newton's Laws of motion, but also the more recently discovered Maxwell's Equations describing electric and magnetic fields, which predict that light always travels at a speed c, equal to about 186,300 miles per second. It follows that any measurement of the speed of any flash of light by an observer in an inertial frame will give the answer c.

We have already noted one counter-intuitive consequence of this, that two different observers moving relative to each other, each measuring the speed of the *same* blob of light relative to himself, will *both* get *c*, even if their relative motion is in the same direction as the motion of the blob of light.

A SIMPLE BUT RELIABLE CLOCK

We mentioned earlier that each of our (inertial) frames of reference is calibrated (had marks at regular intervals along the walls) to measure distances, and has a clock to measure time. Let us now get more specific about the clock—we want one that is easy to understand in any frame of reference. Instead of a pendulum swinging back and forth, which wouldn't work away from the earth's surface anyway, we have a blip of light bouncing back and forth between two mirrors facing each other.

We call this device a *light clock*. To really use it as a timing device we need some way to count the bounces, so we position a photocell at the upper mirror, so that it catches the edge of the blip of light. The photocell clicks when the light hits it, and this regular series of clicks drives the clock hand around, just as for an ordinary clock.

Of course, driving the photocell will eventually use up the blip of light, so we also need some provision to reinforce the blip occasionally, such as a strobe light set to flash just as it passes and thus add to the intensity of the light. Admittedly, this may not be an easy way to build a clock, but the basic idea is simple.

It's easy to figure out how frequently our light clock clicks. If the two mirrors are a distance w apart, the round trip distance for the blip from the photocell mirror to the other mirror and back is $2w$. Since we know the blip always travels at c, we find the round trip time to be $2w/c$, so this is the time between clicks.

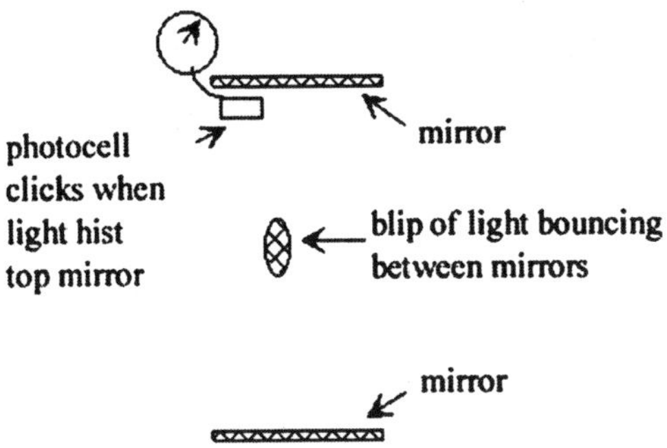

This isn't a very long time for a reasonable sized clock! The crystal in a quartz watch "clicks " of the order of 10,000 times a second. That would correspond to mirrors about nine miles apart, so we need our clock to click about 1,000 times faster than that to get to a reasonable size. Anyway, let us assume that such purely technical problems have been solved.

LOOKING AT SOMEBODY ELSE'S CLOCK

Let us now consider two observers, Jack and Jill, each equipped with a calibrated inertial frame of reference, and a light clock. To be specific, imagine Jack standing on the ground with his light clock next to a straight railroad line, while Jill and her clock are on a large flatbed railroad wagon which is moving down the track at a constant speed v. Jack now decides to check Jill's light clock against his own. He knows the time for his clock is $2w/c$ between clicks. Imagine it to be a slightly misty day, so with binoculars he can actually see the blip of light bouncing between the mirrors of Jill's clock.

How long does he think that blip takes to make a round trip? The one thing he's sure of is that it must be moving at c = 186,300 miles per second, relative to him — that's what Einstein tells him.

So to find the round trip time, all he needs is the round trip distance. This will *not* be $2w$, because the mirrors are on the flatbed wagon moving down the track, so, relative to Jack on the ground, when the blip gets back to the top mirror, that mirror has moved down the track some since the blip left, so the blip actually follows a zigzag path as seen from the ground.

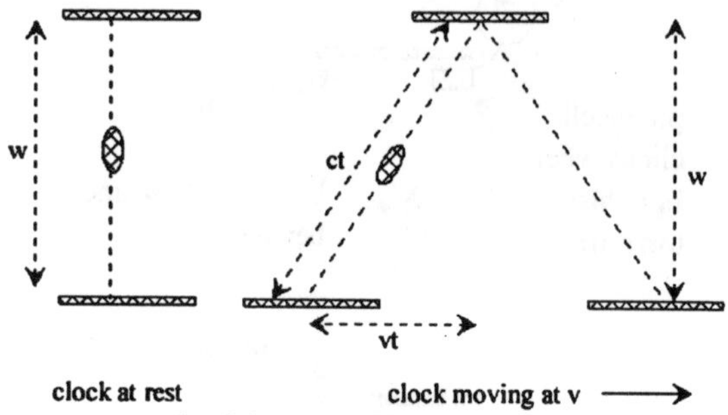

Suppose now the blip in Jill's clock on the moving flatbed wagon takes time t to get from the bottom mirror to the top mirror as measured by Jack standing by the track. Then the length of the "zig" from the bottom mirror to the top mirror is necessarily ct, since that is the distance covered by any blip of light in time t. Meanwhile, the wagon has moved down the track a distance vt, where v is the speed of the wagon. This should begin to look familiar—it is precisely the same as the problem of the swimmer who swims at speed c relative to the water crossing a river flowing at v! We have again a right-angled triangle with hypotenuse ct, and shorter sides vt and w.

From Pythagoras, then,

$$c^2t^2 = v^2t^2 + w^2$$

so

$$t^2(c^2 - v^2) = w^2$$

or

$$t^2(1 - v^2/c^2) = w^2/c^2$$

and, taking the square root of each side, then doubling to get the round trip time

$$2w/c.1/\text{sqrt}(1-v^2/c^2)$$

The effect is not dramatic at real railroad speeds. The correction factor is sqrt($1-v^2/c^2$), which differs from 1 by about one part in a trillion even for a bullet train. It is important to realise that the only reason we chose a *light* clock, as opposed to some other kind of clock, is that its motion is very easy to analyze from a different frame. Jill could have a collection of clocks on the wagon, and would synchronize them all. For example, she could hang her wristwatch right next to the face of the light clock, and observe them together to be sure they always showed the same time. Remember, in her frame her light clock clicks every $2w/c$ seconds, as it is designed to do. Observing this scene from his position beside the track, Jack will see the synchronized light clock and wristwatch next to each other, and, of course, note that the wristwatch is *also* running slow by the factor sqrt($1-v^2/c^2$). In fact, *all* her clocks, including her pulse, are slowed down by this factor according to Jack. Jill is aging more slowly because she's moving.

But this isn't the whole story—we must now turn everything around and look at it from Jill's point of view. Her inertial frame of reference is just as good as Jack's. She sees his light clock to be moving at speed v (backwards) so from her point of view *his* light blip takes the longer zigzag path, which means *his clock runs slower than hers*. That is to say, each of them will see the other to have slower clocks, and be aging more slowly.

This phenomenon is called *time dilation*. It has been verified in recent years by flying very accurate clocks around the world on jetliners and finding they register less time, by the predicted amount, than identical clocks left on the ground.

FITZGERALD CONTRACTION

Consider now the following puzzle: suppose Jill's clock is equipped with a device that stamps a notch on the track once a second. How far apart are the notches? From Jill's point of view, this is pretty easy to answer. She sees the track passing under the wagon at v feet per second, say, to choose convenient units, so the notches will of course be v feet apart. But Jack sees things differently.

He sees Jill's clocks to be running slow, so he will see the notches to be stamped on the track at intervals of 1/sqrt($1-v^2/c^2$) seconds (so for a relativistic train going at $v = 0.8c$, the notches are stamped at intervals of $5/3 = 1.67$ seconds). Since Jack agrees with Jill that the relative speed of the wagon and the track is v, he will assert the notches are not v feet apart, but $v/\text{sqrt}(1-v^2/c^2)$ feet apart, a greater distance. Who is right? It turns out that Jack is right, because the notches are in his frame of reference, so he can wander over to them with a tape measure or whatever, and check the distance. This implies that as a result of her motion, Jill observes the notches to be closer together by a factor $\text{sqrt}(1-v^2/c^2)$ than they would be at rest. This is called the Fitzgerald contraction, and applies not just to the notches, but also to the track and to Jack — everything looks somewhat squashed in the direction of motion!

EXPERIMENTAL EVIDENCE FOR TIME DILATION

The first clear example of time dilation was provided over fifty years ago by an experiment detecting *muons*. These particles are produced at the outer edge of our atmosphere by incoming cosmic rays hitting the first traces of air.

They are unstable particles, with a "half-life" of 1.5 microseconds (1.5 millionths of a second), which means that if at a given time you have 100 of them, 1.5 microseconds later you will have about 50, 1.5 microseconds after that 25, and so on.

Anyway, they are constantly being produced many miles up, and there is a constant rain of them towards the surface of the earth, moving at very close to the speed of light. In 1941, a detector placed near the top of Mount Washington (at 6000 feet above sea level) measured about 570 muons per hour coming in. Now these muons are raining down from above, but dying as they fall, so if we move the detector to a lower altitude we expect it to detect fewer muons because a fraction of those that came down past the 6000 foot level will die before they get to a lower altitude detector. Approximating their speed by that of light, they are raining down at 186,300 miles per second, which turns out to be, conveniently, about 1,000 feet per microsecond.

Thus they should reach the 4500 foot level 1.5 microseconds after passing the 6000 foot level, so, if half of them die off in 1.5 microseconds, as claimed above, we should only expect to register about 570/2 = 285 per hour with the same detector at this level. Dropping another 1500 feet, to the 3000 foot level, we expect about 280/2 = 140 per hour, at 1500 feet about 70 per hour, and at ground level about 35 per hour.

To summarize: given the known rate at which these raining-down unstable muons decay, and given that 570 per hour hit a detector near the top of Mount Washington, we only expect about 35 per hour to survive down to sea level. In fact, when the detector was brought down to sea level, it detected about 400 per hour! How did they survive? The reason they didn't decay is that *in their frame of reference, much less time had passed*. Their actual speed is about 0.994*c*, corresponding to a time dilation factor of about 9, so in the 6 microsecond trip from the top of Mount Washington to sea level, their clocks register only 6/9 – 0.67 microseconds. In this period of time, only about one-quarter of them decay.

What does this look like from the muon's point of view? How do they manage to get so far in so little time? To them, Mount Washington and the earth's surface are approaching at 0.994c, or about 1,000 feet per microsecond. But in the 0.67 microseconds it takes them to get to sea level, it would seem that to them sea level could only get 670 feet closer, so how could they travel the whole 6000 feet from the top of Mount Washington?

The answer is the Fitzgerald contraction—to them Mount Washington is squashed in a vertical direction (the direction of motion) by a factor of $1/\mathrm{sqrt}(1-v^2/c^2)$, the same as the time dilation factor, which for the muons is 9. So, to the muons, Mount Washington is only 670 feet high—this is why they can get down it so fast!

SYNCHRONIZING CLOCKS

Suppose we want to synchronize two clocks that are some distance apart.

We could stand beside one of them and look at the other through a telescope, but we'd have to remember in that case that we are seeing the clock *as it was when the light left it*, and correct accordingly.

Another way to be sure the clocks are synchronized,

assuming they are both accurate, is to start them together. How can we do that? We could, for example, attach a photocell to each clock, so when a flash of light reaches the clock, it begins running. Wagon takes time t to get from the bottom

If, then, we place a flashbulb at the midpoint of the line joining the two clocks, and flash it, the light flash will take the same time to reach the two clocks, so they will start at the same time, and therefore be synchronized.

Let us now put this whole arrangement - the two clocks and the midpoint flashbulb - on a train, and we suppose the train is moving at some speed v to the right, say half the speed of light or so.

Let's look carefully at the clock-synchronizing operation as seen from the ground. In fact, an observer on the ground would say the clocks are *not* synchronized by this operation! The basic reason is that he would see the flash of light from the middle of the train traveling at *c relative to the ground* in each direction, but he would also observe the back of the train coming at v to meet the flash, whereas the front is moving at v away from the bulb, so the light flash must go further to catch up.

In fact, it is not difficult to figure out how much later the flash reaches the front of the train compared with the back of the train, as viewed from the ground. First recall that as viewed from the ground the train has length $L\sqrt{1-\frac{v^2}{c^2}}$..

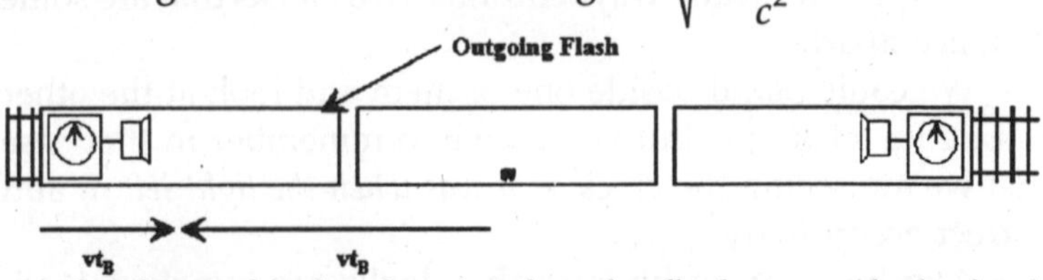

Letting t_B be the time it takes the flash to reach the back

of the train, it is clear from the figure that

$$vt_B + ct_B = \frac{1}{2}L\sqrt{1-\frac{v^2}{c^2}}.$$

from which t_B is given by

$$t_B = \frac{1}{c+v}\frac{1}{2}L\sqrt{1-\frac{v^2}{c^2}}..$$

In a similar way, the time for the flash of light to reach the front of the train is (as measured by a ground observer)

$$t_F = \frac{1}{c-v}\frac{1}{2}L\sqrt{1-\frac{v^2}{c^2}}.$$

Therefore the time difference between the starting of the two clocks, as seen from the ground, is

$$t_F - t_B = \left(\frac{1}{c-v} - \frac{1}{c+v}\right)\frac{1}{2}L\sqrt{1-\frac{v^2}{c^2}}.$$

$$= \frac{2v}{c^2-v^2}\frac{1}{2}L\sqrt{1-\frac{v^2}{c^2}}$$

$$= \frac{2v}{c^2}\frac{1}{1-\frac{v^2}{c^2}}\frac{1}{2}L\sqrt{1-\frac{v^2}{c^2}}$$

$$= \frac{vL}{c^2}\frac{1}{\sqrt{1-\frac{v^2}{c^2}}}$$

Remember, this is the time difference between the starting of the train's back clock and its front clock as measured by an observer on the ground with clocks on the ground. However, to this observer the clocks on the train appear to tick more slowly, by the factor $\sqrt{1-\frac{v^2}{c^2}}$, so that although the ground observer measures the time interval between the starting of the clock at the back of the train and the clock at the front as $\frac{vL}{c^2}\frac{1}{\sqrt{1-\frac{v^2}{c^2}}}$ seconds, he also sees the slow running clock at the back actually reading $\frac{vL}{c^2}$ seconds at the instant he sees the front clock to start.

To summarize: as seen from the ground, the two clocks on the train (which is moving at v in the x-direction) are running slowly, registering only seconds for each second that passes. Equally important, the clocks-which are synchronized by an observer on the train-appear unsynchronized when viewed from the ground, the one at the back of the train reading seconds ahead of the clock at the front of the train, where L is the rest length of the train (the length as measured by an observer on the train).

Note that if $L = 0$, that is, if the clocks are together, both the observers on the train and those on the ground will agree that they are synchronized. We need a *distance* between the clocks, as well as relative motion, to get a disagreement about synchronization.

THE LORENTZ TRANSFORMATIONS

We have already seen that Newtonian mechanics is invariant under the *Galilean* transformations relating two inertial frames moving with relative speed v in the x-direction,

$$x = x' + vt'$$
$$y = y'$$
$$z = z'$$
$$t = t'$$

However, these transformations presuppose that *time* is a well-defined universal concept, that is to say, it's the same time everywhere, and all observers can agree on what time it is. Once we accept the basic postulate of special relativity, however, that the laws of physics, including

Maxwell's equations, are the same in all inertial frames of reference, and consequently the speed of light has the same value in all inertial frames, observers in different frames do *not* agree on whether clocks some distance apart are synchronized.

Furthermore, as we have discussed, measurements of moving objects are compressed in the direction of motion by the Lorentz-Fitzgerald contraction effect. Obviously, the above equations are too naïve! We must think more carefully about time and distance measurement, and construct new transformation equations consistent with special relativity.

Our aim here, then, is to find a set of equations analogous to those above giving the coordinates of an *event* (x,y,z,t) in frame S, for example, a small bomb explosion , as functions of the coordinates (x,y,z,t) of the same event measured in the parallel frame S' which is moving at speed v along the x-axis of frame S. Observers O at the origin in frame S and O' at the origin in frame S synchronize their clocks at $t = t' = 0$, at the instant they pass each other, that is, when the two frames coincide. (Using our previous notation, O is Jack and O' is Jill.)

To determine the time t' at which the bomb exploded in her frame, O' could determine the distance of the point (x',y',z') from the origin, and hence how long it would take light from the explosion to reach her at the origin.

A more direct approach (which is helpful in considering transformations between different frames) is to imagine O' to have a multitude of helpers, with an array of clocks throughout the frame, which have all been synchronized by midpoint flashes as described in the previous lecture.

Then the event — the bomb explosion — will be close to a clock, and that local clock determines the time t' of the event, so we do not need to worry about timing a light signal.

In frame S', then, O' and her crew have clocks all along the x'-axis (as well as everywhere else) and all synchronized:

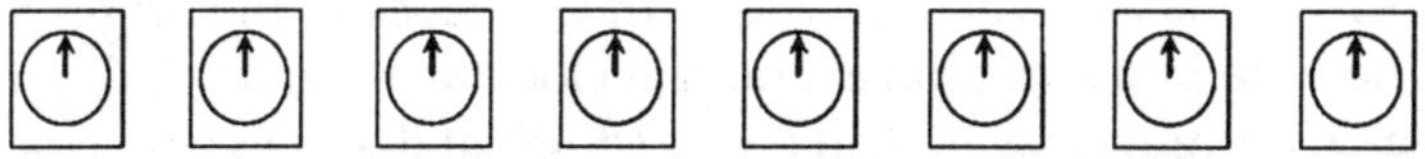

Now consider how this string of clocks appears as viewed by O from frame S. First, since they are all moving at speed v, they will be registering time more slowly by the usual time dilation factor $\sqrt{1 - v^2/c^2}$ than O's own physically identical clocks. Second, they will not be synchronized. From the clocks on a train argument in the last lecture, if the clocks are L apart as measured by O', successive clocks to the right (the direction of motion) will be behind by Lv/c^2 as observed by O.

It should be mentioned that this lack of synchronization as viewed from another frame only occurs for clocks *separated in the direction of relative motion.* Consider two clocks some distance apart on the z' axis of S'. If they are synchronized in S' by both being started by a flash of light from a bulb half way between them, it is clear that as viewed from S the light has to go the same distance to each of the clocks, so they will still be synchronized (although they will start later by the time dilation factor).

Deriving the Lorentz Transformations

Let us now suppose that O' and her crew observe a small bomb to explode in S' at $(x',0,0,t')$. The space coordinates and time (x, y, z, t) of this event as observed by O in the frame S. (S' moves relative to S at speed v along the x-axis). In other words, we shall derive the Lorentz transformations—which are just the equations giving the four coordinates of an event in one inertial frame in terms of the coordinates of the same event in another inertial frame. We take y',z' zero because they transform trivially — there is no Lorentz contraction perpendicular to the motion, so $y = y'$ and $z = z'$.

First, we consider at what *time* the bomb explodes as measured by O. O''s crew found the bomb to explode at time t' as measured by a local clock, that is, one located at the site of the explosion, x'. Now, as observed by O from frame S, O''s clock at x' is *not* synchronized with O''s clock at the origin S'. When the bomb explodes and the clock at x' reads t', O will see O''s origin clock to read $t' + vx'/c^2$. What does O's own clock read at this point? Recall that O, O' synchronized their origin clocks at the moment they were together, at $t = t' = 0$. Subsequently, O will have observed O''s clock to be running slowly by the time-dilation factor.

Therefore, when at the instant of the explosion he sees O''s origin clock to be reading $t' + vx'/c^2$, he will find that the true time t in his frame is equal to this appropriately scaled to allow for time dilation, that is,

$$t = \frac{t' + vx'/c^2}{\sqrt{1 - v^2/c^2}}$$

This is the first of the Lorentz transformations.

The second question is: *where* does O observe the explosion to occur?

Since it occurs at time t after O' passed O, O' is vt meters beyond O at the time of the explosion. The explosion takes place x' meters beyond O', as measured by O', but of course O will see that distance x' as contracted to $x'\sqrt{1 - v^2/c^2}$ since it's in a moving frame.

Therefore O observes the explosion at point x given by

$$x = vt + x'\sqrt{1 - v^2/c^2}$$

This can be written as an equation for x in terms of x', t' by substituting for t using the first Lorentz transformation above, to give

$$x = \frac{x' = vt'}{\sqrt{1 - v^2/c^2}}$$

Therefore, we have found the Lorentz transformations expressing the coordinates (x,y,z,t) of an event in frame S in terms of the coordinates (x',y',z',t') of the same event in frame S':

$$x = \frac{x' = vt'}{\sqrt{1 - v^2/c^2}}$$

$$y = y'$$
$$z = z'$$

$$t = \frac{t' + vx'/c'}{\sqrt{1 - v^2/c^2}}$$

Notice that nothing in the above derivation depends on the x-velocity v of S' relative to S being positive. Therefore, the inverse transformation (from (x,y,z,t) to (x', y', z',t')) has exactly the same form as that given above with v replaced by $-v$.

SPHERES OF LIGHT

Consider now the following scenario: suppose that as O' passes O (the instant both of them agree is at time $t' = t = 0$) O' flashes a bright light, which she observes to create an expanding spherical shell of light, centered on herself (imagine it's a slightly foggy day, so she can see how the ripple of light travels outwards).

At time t', then O' (or, to be precise, her local observers out there in the frame) will see a shell of light of radius ct', that is to say, they will see the light to have reached all points (x', y', z') on the surface

$$x'^2 + y'^2 + z'^2 = c^2t'^2$$

Question: how do O and his observers stationed throughout the frame S see this light as rippling outwards?

To answer this question, notice that the above equation for where the light is in frame S' at a particular time t' can be written

$$x'^2 + y'^2 + z'^2 - c^2t'^2 = 0,$$

and can be thought of as *a surface in the four dimensional* (x', y', z', t') *space,* the totality of all the "events" of the light reaching any particular point.

Now, to find the corresponding surface of events in the four dimensional (x, y, z, t) space, all we have to do is to change from one set of variables to the other using the Lorentz transformations:

$$x' = \frac{x = vt}{\sqrt{1 - v^2/c^2}}$$

$$y' = y$$

$$z' = z$$

$$t = \frac{t' + vx'/c'}{\sqrt{1 - v^2/c^2}}$$

On putting these values,

of x', y', z', t' into $x'^2 + y'^2 + z'^2 - c^2t'^2 = 0$, we find that the corresponding surface of events in (x, y, z, t) space is:

$$x^2 + y^2 + z^2 - c^2t^2 = 0$$

This means that at time t, O and his observers in frame S will say the light has reached a spherical surface centered on O.

How can O' and O, as they move further apart, possibly both be right in maintaining that at any given instant the outward moving light pulse has a spherical shape, each saying it is centered on herself or himself?

Imagine the light shell as O' sees it — at the instant t' she sees a sphere of radius r', in particular she sees the light to have reached the spots $+r'$ and $-r'$ on the x' axis. But from O's point of view the expanding light sphere does *not* reach the point $+r'$ at the same time it reaches $-r'$! (This is just the old story of synchronizing the two clocks at the front and back of the train one more time).

That is why O does not see O''s sphere: the arrival of the light at the sphere of radius r' around O' at time t' corresponds in S to a continuum of different events happening at different times.

Lorentz Invariants

We found above that for an event (x', y', z', t') for which $x'^2 + y'^2 + z'^2 - c^2t'^2 = 0$, the coordinates of the event (x, y, z, t) as measured in the other frame S satisfy $x^2 + y^2 + z^2 - c^2t^2 = 0$. The quantity $x^2 + y^2 + z^2 - c^2t^2$ is said to be a *Lorentz invariant*, it doesn't vary on going from one frame to another.

A simple two-dimensional analogy to this invariant is given by considering two sets of axes, Oxy and $Ox'y'$ having the same origin O, but the axis Ox' is at an angle to Ox, so one set of axes is the same as the other set but rotated. The point P with coordinates (x,y) has coordinates (x', y') measured on the $Ox'y'$ axes.

The square of the distance of the point P from the common origin O is $x^2 + y^2$ and is also $x'^2 + y'^2$, so for the transformation from coordinates (x,y) to (x', y'), $x^2 + y^2$ is an invariant. Similarly, if a point P_1 has coordinates (x_1,y_1) and (x_1', y_1') and P_2 has coordinates (x_2,y_2) and (x_2',y_2') then clearly the two points are the same distance apart as measured with respect to the two sets of axes, so

$$(x_2 - x_2)^2 + y_1 - y_2)^2 = (x'_2 - x'_2)^2 + (y'_1 - y'_2)^2$$

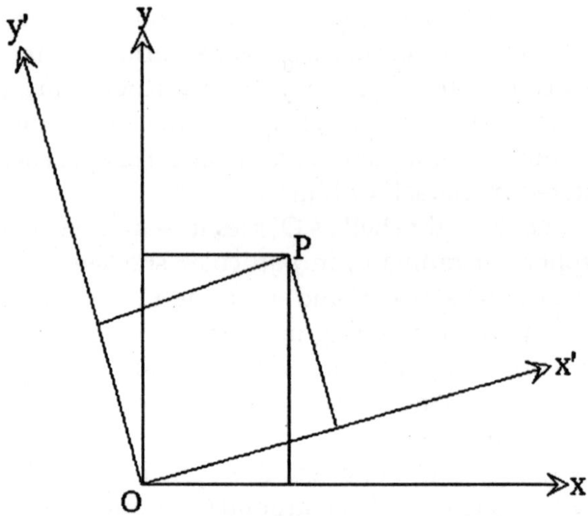

This is really obvious-the distance between two points in an ordinary plane can't depend on the angle at which we choose to set our coordinate axes.

The Lorentz analog of this, dropping the *y,z* coordinates, can be written

$$c^2(t_1 - t_2)^2 - (x_1 - x_2)^2 = c^2(t'_1 - t'_2)^2 - (x'_1 - x'_2)^2 = s^2,$$

say, where *s* is some sort of measure of the "distance" between the two events (x_1,t_1) and (x_2, t_2).

This s is sometimes called the "space-time interval". The big difference from the two-dimensional rotation case is that s^2 can be positive or negative. For s^2 negative, the "distance" itself would be imaginary, so obviously this is rather a formal concept In practice it's clear enough what to do in any particular case — the cases of spacelike and timelike separated events are best dealt with separately, at least to begin with.

Consider first two events simultaneous in frame *S'*, so $t_1' = t_2'$. They will not be simultaneous in frame *S*, but they will satisfy

$$(x_2 - x_2)^2 - c^2 (t_1 - t_2)^2 \geq 0.$$

We say the two events are *spacelike separated*. This means that they are sufficiently removed spatially that a light signal could not have time between them to get from one to the other, so one of these events could not be the *cause* of the other. The sequence of two events can be different in different frames if the events are spacelike separated. Consider again the starting of the two clocks at the front and back of a train as seen from the ground: the back clock starts first.

Now imagine viewing this from a faster train overtaking the clock train — now the front clock will be the first to start. The important point is that although these events appear to occur in a different order in a different frame, neither of them could be the cause of the other, so cause and effect are not switched around.

Consider now two events which occur at the same place in frame S' at different times, (x'_1, t'_1) and (x'_2, t'_1). Then in frame S:

$$c^2(t_1 - t_2)^2 - (x_1 - x_2)^2 = c^2(t'_1 - t') > 0$$

These events are said to be *timelike* separated. There is no frame in which they are simultaneous. "Cause and effect" events are timelike separated.

The Light Cone

Let us try to visualize the surface in four-dimensional space described by the outgoing shell of light from a single flash,

$$x^2 + y^2 + z^2 - c^2t^2 = 0.$$

It is helpful to think about a simpler situation, the circular ripple spreading on the surface of calm water from a pebble falling in. Taking c here to be the speed of the water waves, it is easy to see that at time t after the splash the ripple is at

$$x^2 + y^2 - c^2t^2 = 0..$$

Now think about this as a surface in the three-dimensional space (x, y, t). The plane corresponding to time t cuts this surface in a circle of radius ct. This means the surface is a cone with its point at the origin. The four-dimensional space flash-of-light surface is not so easy to visualize, but is clearly the higher-dimensional analog: the plane surface corresponding to time t cuts it in a sphere instead of a circle. This surface is called the *lightcone*.

We have stated above that the separation of a point $P(x, y, z, t)$ from the origin is spacelike if $x^2 + y^2 + z^2 - c^2t^2 > 0$.

and timelike if $x^2 + y^2 + z^2 - c^2t^2 < 0$. It is said to be *lightlike* if $x^2 + y^2 + z^2 - c^2t^2 = 0$. Points on the light cone described above are lightlike separated from the origin. To be precise, the points corresponding to an outgoing shell of light from a flash at the origin at $t = 0$ form the *forward light cone*. Since the equation depends only on t^2, there is a solution with t negative, the "backward light cone", just the reflection of the forward light cone in the plane $t = 0$.

Possible causal connections are as follows: an event at the origin (0, 0, 0, 0) could cause an event in the forward light cone: so that is the "future", as seen from the origin. Events in the backward light cone — the "past" — could cause an event at the origin. There can be no causal link between an event at the origin and an event outside the light cones, since the separation is spacelike: outside the light cones is "elsewhere" as viewed from the origin.

EINSTEIN'S DEFINITION

As you can see from the lectures so far, although Einstein's theory of special relativity solves the problem posed by the Michelson-Morley experiment — the nonexistence of an ether — it is at a price. The simple assertion that the speed of a flash of light is always c in any inertial frame leads to consequences that defy common sense. When this was pointed out somewhat forcefully to Einstein, his response was that common sense is the layer of prejudices put down before the age of eighteen. All our intuition about space, time and motion is based on childhood observation of a world in which no objects move at speeds comparable to that of light. Perhaps if we had been raised in a civilization zipping around the universe in spaceships moving at relativistic speeds, Einstein's assertions about space and time would just seem to be common sense.

The real question, from a scientific point of view, is not whether special relativity defies common sense, but whether it can be shown to lead to a *contradiction*. If that is so, common sense wins. Ever since the theory was published, people have been writing papers claiming it *does* lead to contradictions. The previous lecture, the worked example on time dilation, shows how careful analysis of an apparent contradiction leads to the conclusion that in fact there was no contradiction after all. In this lecture, we shall consider other apparent contradictions and think about how to resolve them. This is the best way to build up an understanding of relativity.

TRAPPING A TRAIN IN A TUNNEL

One of the first paradoxes to be aired was based on the Fitzgerald contraction. Recall that any object moving relative to an observer will be seen by that observer to be contracted, foreshortened in the direction of motion by the ubiquitous factor sqrt(1 - v^2/c^2). Einstein lived in Switzerland, a very mountainous country where the railroads between towns often go through tunnels deep in the mountains.

Suppose a train of length L is moving along a straight track at a relativistic speed and enters a tunnel, also of length L. There are bandits inhabiting the mountain above the tunnel. They observe a short train, one of length Lsqrt(1 - v^2/c^2), so they wait until this short train is completely inside the tunnel of length L, then they close doors at the two ends, and the train is trapped fully inside the mountain. Now look at this same scenario from the point of view of someone on the train. He sees a train of length L, approaching a tunnel of length Lsqrt(1 - v^2/c^2), so the tunnel is not as long as the train from his viewpoint! What does he think happens when the bandits close both the doors?

The Tunnel Doors are Closed Simultaneously

The key to understanding what is happening here is that we said the bandits closed the two doors at the ends of the tunnel *at the same time*. How could they arrange to do that, since the doors are far apart? They could use walkie-talkies, which transmit radio waves, or just flash a light down the tunnel, since it's long and straight.

Remember, though, that the train is itself going at a speed close to that of light, so they have to be quite precise about this timing! The simplest way to imaging them synchronizing the closings of the two doors is to assume they know the train's timetable, and at a prearranged appropriate time, a light is flashed halfway down the tunnel, and the end doors are closed when the flash of light reaches the ends of the tunnel. Assuming the light was positioned correctly in the middle of the tunnel, that should ensure that the two doors close simultaneously.

Now consider this door-closing operation from the point of view of someone on the train. Assume he's in an observation car and has incredible eyesight, and there's a little mist, so he actually sees the light flash, and the two flashes traveling down the tunnels towards the two end doors. Of course, *the train is a perfectly good inertial frame,* so he sees these two flashes to be traveling in opposite directions, but *both at c, relative to the train.* Meanwhile, he sees the tunnel itself to be moving rapidly relative to the train. Let us say the train enters the mountain through the "front" door.

The observer will see the door at the other end of the tunnel, the "back" door, to be rushing towards him, and rushing to meet the flash of light. Meanwhile, once he's in the tunnel, the front door is receding rapidly behind him, so the flash of light making its way to that door has to travel further to catch it. So the two flashes of light going down the tunnel in opposite directions do not reach the two doors simultaneously as seen from the train.

The concept of simultaneity, events happening at the same time, is not *invariant as we move from one inertial frame to another.* The man on the train sees the back door close first, and, if it is not quickly reopened, the front of the train will pile into it before the front door is closed behind the train.

FITZGERALD CONTRACTION

The above discussion is based on Einstein's prediction that objects moving at relativistic speed appear shrunken in their direction of motion. How do we know that they're not shrunken in all three directions, i.e. moving objects maybe keep the same shape, but just get smaller? This can be seen *not* to be the case through a symmetry argument, also due to Einstein. Suppose two trains traveling at equal and opposite relativistic speeds, one north, one south, pass on parallel tracks. Suppose two passengers of equal height, one on each train, are standing leaning slightly out of open windows so that their noses should very lightly touch as they pass each other. Now, if *N* (the northbound passenger) sees *S* as shrunken in height, *N*'s nose will brush against *S*'s forehead, say, and *N* will feel *S*'s nose brush his chin.

Afterwards, then, *N* will have a bruised chin (plus nose), *S* a bruised forehead (plus nose). But this is a perfectly symmetric problem, so *S* would say *N* had the bruised forehead, etc. They can both get off their trains at the next stations and get together to check out bruises. They must certainly be symmetrical! The only *consistent symmetrical solution* is given by asserting that *neither* sees the other to shrink in height (i.e. in the direction perpendicular to their relative motion), so that their noses touch each other. Therefore, the Lorentz contraction *only* operates in the direction of motion, objects get squashed but not shrunken.

Perhaps the most famous of the paradoxes of special relativity, which was still being hotly debated in national journals in the fifties, is the twin paradox. The scenario is as follows. One of two twins - the sister — is an astronaut. (Flouting tradition, we will take fraternal rather than identical twins, so that we can use "he" and "she" to make clear which twin we mean).

She sets off in a relativistic spaceship to alpha-centauri, four light-years away, at a speed of, say, 0.6*c*. When she gets there, she immediately turns around and comes back. As seen by her brother on earth, her clocks ran slowly by the time dilation factor sqrt($1-v^2/c^2$), so although the round trip took 8/0.6 years = 160 months by earth time, she has only aged by 4/5 of that, or 128 months . So as she steps down out of the spaceship, she is 32 months younger than her twin brother.

But wait a minute — how does this look from *her* point of view? She sees the earth to be moving at 0.6*c*, first away from her then towards her. So she must see her brother's clock on earth to be running slow! So doesn't she expect her brother on earth to be the younger one after this trip?

The key to this paradox is that this situation is not as symmetrical as it looks. The two twins have quite different experiences. The one on the spaceship is *not* in an inertial frame during the initial acceleration *and* the turnaround and braking periods. (To get an idea of the speeds involved, to get to 0.6*c* at the acceleration of a falling stone would take over six months.) Our analysis of how a clock in one inertial frame looks as viewed from another doesn't work during times when one of the frames isn't inertial - in other words, when one is accelerating.

THE TWINS STAY IN TOUCH

To try to see just how the difference in ages might develop, let us imagine that the twins stay in touch with each other throughout the trip. Each twin flashes a powerful light once a month, according to their calendars and clocks, so that by counting the flashes, each one can monitor how fast the other one is aging.

FIGURING THE OBSERVED TIME BETWEEN FLASHES

In some ways, that was a very small scale version of the present problem. Recall that we had two "ground" clocks only one million miles apart. As the astronaut, conveniently moving at 0.6*c*, passed the first ground clock, both that clock and her own clock read zero. As she passed the second ground clock, her own clock read 8 seconds and the *first* ground clock, which she photographed at that instant, she observed to read 4 seconds.

That is to say, after 8 seconds had elapsed on her own clock, constant *observation* of the first ground clock would have revealed it to have registered only 4 seconds. (This effect is compounded of time dilation and the fact that as she moves away, the light from the clock is taking longer and longer to reach her.)

Our twin problem is the same thing, at the same speed, but over a longer time - we conclude that observation of any earth clock from the receding spacecraft will reveal it to be running *at half speed,* so the brother's flashes will be seen at the spacecraft to arrive every two months, by spacecraft time.

Symmetrically, as long as the brother on earth observes his sister's spacecraft to be moving away at 0.6*c*, he will see light from her flashes to be arriving at the earth every two months by earth time. To figure the frequency of her brother's flashes observed as she returns towards earth, we have to go back to our previous example and find how the astronaut traveling at 0.6*c* observes time to be registered by the *second* ground clock, the one she's approaching.

We know that as she passes that clock, it reads 10 seconds and her own clock reads 8 seconds. We must figure out what she would have seen that second ground clock to read had she glanced at it through a telescope as she passed the first ground clock, at which point both her own clock and the first ground clock read zero. But at that instant, the reading she would see on the second ground clock must be the same as would be seen by an observer on the ground, standing by the first ground clock and observing the second ground clock through a telescope. Since the ground observer knows both ground clocks are synchronized, and the first ground clock reads zero, and the second is 6 light seconds distant, it must read -6 seconds if observed at that instant.

Hence the astronaut will observe the second ground clock to progress from -6 seconds to +10 seconds during the period that her own clock goes from 0 to 8 seconds. In other words, she sees the clock she is approaching at 0.6*c* to be running *at double speed.* Finally, back to the twins. During her journey back to earth, the sister will see the brother's light flashing twice a month.

At 0.6*c*, she sees the distance to alpha-centauri to be contracted by the familiar sqrt(1 - v^2/c^2) = 0.8 to a distance of 3.2 light years, which at 0.6*c* will take her a time 5.333 years, or, more conveniently, 64 months. During the outward trip, then, she will see 32 flashes from home, she will see her brother to age by 32 months. Her return trip will also take 64 months, during which time she will see 128 flashes, so over the whole trip she will see 128 + 32 = 160 flashes, so she will have seen her brother to age by 160 months or 13 years 4 months. At first glance, this sounds the same as what she sees — but it isn't! The important question to ask is *when* does he see her turn around? To him, her outward journey of 4 light years' distance at a speed of 0.6*c* takes her 4/0.6 years, or 80 months. BUT he doesn't *see* her turn around until 4 years later, because of the time light takes to get back to earth from alpha-centauri! In other words, he will actually see her aging at half his rate for 80 + 48 = 128 months, during which time he will see 64 flashes.

When he *sees* his sister turn around, she is already more than half way back! Remember, in his frame the whole trip takes 160 months (8 light years at 0.6*c*) so he will only see her aging at twice his rate during the last 160 - 128 = 32 months, during which period he will see all 64 flashes she sent out on her return trip.

Therefore, by counting the flashes of light she transmitted once a month, he will conclude she has aged 128 months on the trip, which by his clock and calendar took 160 months. So when she steps off the spacecraft 32 months younger than her twin brother, neither of them will be surprised!

THE DOPPLER EFFECT

The above analysis hinges on the fact that a traveller approaching a flashing light at 0.6*c* will see it flashing at *double* its "natural" rate - the rate observed by someone standing still with the light - and a traveller receding at 0.6*c* from a flashing light will see it to flash at only *half* its natural rate.

This is a particular example of the *Doppler Effect*, first discussed in 1842 by the German physicist Christian Doppler. There is a Doppler Effect for sound waves too. Sound is generated by a vibrating object sending a succession of pressure pulses through the air. These pressure waves are analogous to the flashes of light. If you are approaching a sound source you will encounter the pressure waves more frequently than if you stand still.

This means you will hear a higher frequency sound. If the distance between you and the source of sound is increasing, you will hear a lower frequency. This is why the note of a jet plane or a siren goes lower as it passes you. The details of the Doppler Effect for sound are a little different than those for light, because the speed of sound is not the same for all observers - it's 330 meters per second relative to the air.

It isn't difficult to find the general formula for the Doppler shift, that is, the change in frequency observed when the source of waves (or periodic signals in general) is moving. For example, consider a light flashing once a second as observed by someone in the same frame as the light. Let us imagine the light to be attached to a spaceship, passing us at a relativistic speed v, and imagine we see a flash at the instant it passes us. When will we see the next flash?

First, the spaceship's clock is running slow according to us, so it will take $1/\text{sqrt}(1 - v^2/c^2)$ seconds before it emits the next flash. But it's also moving away from us, so we won't see that next flash until the light has traveled back to us over the distance covered by the spaceship between flashes. Since the spaceship is traveling at v, and the time between flashes as measured in our frame is $1/\text{sqrt}(1 - v^2/c^2)$, the distance the spaceship covers between flashes measures in our frame is $v/\text{sqrt}(1 - v^2/c^2)$. Since the light coming back to us from the flash is traveling at c, it covers this distance in time $v/c\text{sqrt}(1 - v^2/c^2)$.

Thus the total time between our observing the first flash as the spaceship passes close by us and the second flash emitted one second later by the spaceship clock is:

$$1/\text{sqrt}(1 - v^2/c^2) + v/c\text{sqrt}(1 - v^2/c^2) = \text{sqrt}((1 + v/c)/(1 - v/c))$$

The change in frequency is the inverse of the change in time between flashes. Also, notice that the shift for a spaceship *approaching* at speed v is the *inverse* of that for one *receding* at speed v. We already established that above for the special case of $v = 0.6c$, where we found the frequency halved for the spaceship receding, doubled for it approaching. There are some significant differences between the Doppler effect for light and that for sound.

The most obvious one is that if something is approaching you at a speed higher than the speed of sound, you won't hear a thing until it hits. Nothing can approach you at greater than the speed of light. Another difference is that the Doppler shift for light depends only on the velocity of the emitter relative to that of the receiver.

For sound, the shift depends on the velocities of both of them relative to the air. Finally, consider a signal from a moving object moving along a straight line, but not towards you. For example, consider a train moving along a straight track, and the nearest point on the track is one mile from where you're standing.

What frequency do you hear for the signal emitted when the train was at the nearest point to you on the track? For a sound signal, you would hear no Doppler shifting at this point. For a light signal, however, you *would* see a frequency shift downwards equal to the time dilation factor. This is called the *transverse* Doppler shift, and is famous historically because it was first detected in 1938 by two experimenters, Ives and Stillwell, who even then didn't believe in special relativity! They interpreted their result as the slowing down of a clock as it moved through the aether.

An important astronomical application of the Doppler Effect is the *red shift*. The light from very distant galaxies is redder than the light from similar galaxies nearer to us. This is because the further away a galaxy is, the faster it is moving away from us, as the Universe expands. The light is redder because red light is low frequency light (blue is high) and we see low frequency light for the same reason that the astronaut receding from earth sees flashes less frequently. In fact, the farthest away galaxies we can see are receding faster than the 0.6*c* of our astronaut!

RELATIVISTIC DYNAMICS

The first coherent statement of what physicists now call *relativity* was Galileo's observation almost four hundred years ago that if you were in a large closed room, you could not tell by observing how things move-living things, thrown things, dripping liquids-whether the room was at rest in a building, say, or below decks in a large ship moving with a steady velocity. More technically (but really saying the same thing!)

we would put it that *the laws of motion are the same in any inertial frame.* That is, these laws really only describe *relative* positions and velocities. In particular, they do not single out a special inertial frame as the one that's "really at rest". This was later all written down more formally, in terms of *Galilean transformations*.

Using these simple linear equations, motion analyzed in terms of positions and velocities in one inertial frame could be translated into any other. When, after Galileo, Newton wrote down his Three Laws of Motion, they were of course invariant under the Galilean transformations, and valid in any inertial frame.

About two hundred years ago, it became clear that light was not just a stream of particles (as Newton had thought) but manifested definite *wavelike* properties. This led naturally to the question of what, exactly, was waving, and the consensus was that space was filled with an *aether*, and light waves were ripples in this all-pervading aether analogous to sound waves in air. Maxwell's discovery that the equations describing electromagnetic phenomena had wavelike solutions, and predicted a speed which coincided with the measured speed of light, suggested that electric and magnetic fields were stresses or strains in the aether, and Maxwell's equations were presumably only precisely correct in the frame in which the aether was at rest. However, very precise experiments which should have been able to detect this aether all failed.

Almost one hundred years ago, Einstein suggested that maybe *all* the laws of physics were the same in *all* inertial frames, generalizing Galileo's pronouncements concerning motion to include the more recently discovered laws of electricity and magnetism. This would imply there could be no special "really at rest" frame, even for light propagation, and hence no aether.

This is a very appealing and very simple concept: the same laws apply in all frames. What could be more reasonable? Though, it turns out to clash with some beliefs about space and time deeply held by everybody encountering this for the first time. The central prediction is that since the speed of light follows from the laws of physics (Maxwell's equations) and some simple electrostatic and magnetostatic experiments, which are clearly frame-independent, *the speed of light is the same in all inertial frames*. That is to say, the speed of a *particular flash* of light will always be measured to be 3 10^8 meters per second even if measured by different observers moving rapidly relative to each other, where each observer measures the speed of the flash relative to himself. Nevertheless, experiments have show again and again that Einstein's elegant insight is right, and everybody's deeply held beliefs are wrong.

We have discussed in detail the *kinematical* consequences of Einstein's postulate-how measurements of position, time and velocity in one frame relate to those in another, and how apparent paradoxes can be resolved by careful analysis. So far, though, we have not thought much about dynamics.

We know that Newton's Laws of Motion were invariant under the Galilean transformations between inertial frames. We now know that the Galilean transformations are in fact *incorrect* except in the low speed nonrelativistic limit. Therefore, we had better look carefully at Newton's Laws of Motion in light of our new knowledge.

Newton's Laws Revisited

Newton's First Law, the Principle of Inertia, that an object subject to no external forces will continue to move in a straight line at steady speed, is equally valid in special relativity. Indeed, it is the defining property of an inertial frame that this *is* true, and the content of special relativity is transformations between such frames.

Newton's Second Law, stated in the form force = mass x acceleration, cannot be true as it stands in special relativity. This is evident from the formula we derived for addition of velocities. Think of a rocket having many stages, each sufficient to boost the remainder of the rocket (including the unused stages) to $c/2$ from rest. We could fire them one after the other in a carefully timed way to generate a continuous large force on the rocket, which would get it to $c/2$ in the first firing. If the acceleration continued, the rocket would very soon be exceeding the speed of light.

Yet we know from the addition of velocities formula that in fact the rocket never reaches *c*. Evidently, Newton's Second Law needs updating.

Newton's Third Law, action = reaction, also has problems. Consider some attractive force between two rapidly moving bodies. As their distance apart varies, so does the force of attraction. We might be tempted to say that the force of *A* on *B* is the opposite of the force of *B* on *A*, at each instant of time, but that implies *simultaneous* measurements at two bodies some distance from each other, and if it happens to be true in *A*'s inertial frame, it won't be in *B*'s.

Conservation Laws

In nonrelativistic Newtonian physics, the Third Law tells us that two interacting bodies feel equal but opposite forces from the interaction. Therefore from the Second Law, the rate of change of momentum of one of the bodies is equal and opposite to that of the other body, thus the *total* rate of change of momentum of the system caused by the interaction is zero. Consequently, for any *closed* dynamical system (no outside forces acting) *the total momentum never changes.* This is the *law of conservation of momentum.* It does *not* depend on the details of the forces of interaction between the bodies, only that they be equal and opposite.

The other major dynamical conservation law is the conservation of energy. This was not fully formulated until long after Newton, when it became clear that frictional heat generation, for example, could quantitatively account for the apparent loss of kinetic plus potential energy in actual dynamical systems.

Although these conservation laws were originally formulated within a Newtonian worldview, their very general nature suggested to Einstein that they might have a wider validity. Therefore, as a working hypothesis, he assumed them to be satisfied *in all inertial frames,* and explored the consequences. We follow that approach.

MOMENTUM CONSERVATION

As a warm-up exercise, let us consider conservation of momentum for a collision of two balls on a pool table. We draw a chalk line down the middle of the pool table, and shoot the balls close to, but on opposite sides of, the chalk line from either end, at the same speed, so they will hit in the middle with a glancing blow, which will turn their velocities through a small angle. In other words, if initially we say their (equal magnitude, opposite direction) velocities were parallel to the x-direction — the chalk line — then after the collision they will also have equal and opposite small velocities in the y-direction. (The x-direction velocities will have decreased very slightly).

COLLISION

Now let us repeat the exercise on a grand scale. Suppose somewhere in space, far from any gravitational fields, we set out a string one million miles long. (It could be between our two clocks in the time dilation experiment). This string corresponds to the chalk line on the pool table. Suppose now we have two identical spaceships approaching each other with equal and opposite velocities parallel to the string from the two ends of the string, aimed so that they suffer a slight glancing collision when they meet in the middle. It is evident from the symmetry of the situation that momentum is conserved in both directions. In particular, the rate at which one spaceship moves away from the string after the collision - its y-velocity - is equal and opposite to the rate at which the other one moves away from the string.

But now consider this collision as observed by someone in one of the spaceships, call it A. (Remember, momentum must be conserved in all inertial frames — they are all equivalent — there is nothing special about the frame in which the string is at rest).

Before the collision, he sees the string moving very fast by the window, say a few meters away. After the collision, he sees the string to be moving away, at, say, 15 meters per second. This is because spaceship A has picked up a velocity perpendicular to the string of 15 meters per second.

Meanwhile, since this is a completely symmetrical situation, an observer on spaceship B would certainly deduce that her spaceship was moving away from the string at 15 meters per second as well.

The crucial question is: how fast does an observer in spaceship A see spaceship B to be moving away from the string? Let us suppose that relative to spaceship A, spaceship B is moving away (in the x-direction) at 0.6c. First, recall that distances perpendicular to the direction of motion are not Lorentz contracted. Therefore, when the observer in spaceship B says she has moved 15 meters further away from the string in a one second interval, the observer watching this movement from spaceship A will agree on the 15 meters - but disagree on the one second! He will say her clocks run slow, so as measured by his clocks 1.25 seconds will have elapsed as she moves 15 meters in the y-direction.

It follows that, as a result of time dilation, this collision as viewed from spaceship A does not cause equal and opposite velocities for the two spaceships in the y-direction. Initially, both spaceships were moving parallel to the x-axis - there was zero momentum in the y-direction. Consider y-direction momentum conservation in the inertial frame in which A was initially at rest. An observer in that frame measuring y-velocities after the collision will find A to be moving at 15 meters per second, B to be moving at -0.8 x 15 meters per second in the y-direction. So how can we argue there is zero total momentum in the y-direction after the collision, when the identical spaceships do not have equal and opposite velocities?

Einstein Rescues Momentum Conservation

Einstein was so sure that momentum conservation must always hold that he rescued it with a bold hypothesis: the mass of an object must depend on its speed! In fact, the mass must increase with speed in just such a way as to cancel out the lower y-direction velocity resulting from time dilation. That is to say, if an object at rest has a mass M, moving at a speed v it will have a mass $M/\text{sqrt}(1 - v^2/c^2)$. Note that this is an undetectably small effect at ordinary speeds, but as an object approaches the speed of light, the mass increases without limit!

Of course, we have taken a very special case here - a particular kind of collision. The reader might well wonder if the same mass correction would work in other types of collision, for example a straight line collision in which a heavy object rear-ends a lighter object. The algebra is straightforward, if tedious, and it is found that this mass correction factor does indeed ensure momentum conservation for any collision in all inertial frames.

Mass Really Does Increase with Speed

Deciding that masses of objects must depend on speed like this seems a heavy price to pay to rescue conservation of momentum! However, it is a prediction that is not difficult to check by experiment. The first confirmation came in 1908, measuring the mass of fast electrons in a vacuum tube. In fact, the electrons in a colour TV tube are about half a percent heavier than electrons at rest, and this must be allowed for in calculating the magnetic fields used to guide them to the screen.

Much more dramatically, in modern particle accelerators very powerful electric fields are used to accelerate electrons, protons and other particles. It is found in practice that these particles become heavier and heavier as the speed of light is approached, and hence need greater and greater forces for further acceleration. Consequently, the speed of light is a natural absolute speed limit. Particles are accelerated to speeds where their mass is thousands of times greater than their mass measured at rest, usually called the "rest mass".

Mass and Energy Conservation in Special Relativity

As everyone has heard, in special relativity mass and energy are not separately conserved, in certain situations mass m can be converted to energy $E = mc^2$. This equivalence is closely related to the mass increase with speed, as we shall see. In fact, French derives it by analyzing a collision and stipulating mass conservation in the collision, where mass is now the speed-dependent mass discussed above. Part of French's point in doing this is to demonstrate that it follows from very general considerations, for example concepts like force do not need to be introduced. We shall take a less sophisticated route below, in which we find the kinetic energy of an object as it is accelerated in a straight line by a force, such as the electric field acting on an electron in a linear accelerator, and assume Newton's second law is valid provided mass x acceleration is replaced by rate of change of momentum, a more general concept.

KINETIC ENERGY AND MASS FOR VERY FAST PARTICLES

Let's first think about the kinetic energy of a particle traveling close to the speed of light. If a force F acts on the particle accelerating it in its direction of motion, and the particle moves through a distance d in the direction the force is pushing, the force does work Fd, and this must go into the kinetic energy of the particle (assuming as always conservation of energy).

As a warm up, recall the elementary derivation of the kinetic energy $\frac{1}{2}mv^2$ of an ordinary *non-relativistic* (i.e. slow moving) object of mass m. This can be done by accelerating the mass with a constant force F, and finding the work done by the force (force distance) to get it to speed v from a standing start.

The kinetic energy of the mass, $E = \frac{1}{2}mv^2$, equals the work done by the force in bringing the mass up to that speed from rest: if it takes time t, and the (uniform) acceleration is a, then the final speed $v = at$, the average speed during acceleration is $\frac{1}{2}v$, so the distance traveled is $\frac{1}{2}vt$, the work done by the force is $\frac{1}{2}Fvt = \frac{1}{2}mavt = \frac{1}{2}mv^2$. It is easy to extend this to show that if the mass was initially moving at speed u, the work done by the force is $\frac{1}{2}mv^2 - \frac{1}{2}mu^2$.

We are now ready to think about the change in kinetic energy of a particle acted on by an accelerating force when the particle is already moving practically at the speed of light, like particles in modern accelerators. The most straightforward way to see what happens is to use Newton's Second Law, which in the form

Force = rate of change of momentum

is still true, but close to the speed of light the speed changes negligibly as the force continues to work-instead, the mass increases. Therefore to an excellent approximation,

Force = (rate of change of mass) c

where as usual *c* is the speed of light.

To give a concrete example, let us imagine we have a particle moving very close to the speed of light, and we apply an accelerating force *F* as the particle travels in a straight line through a distance *d*. (In a particle accelerator, for example, the particle with charge *q* could enter a region of uniform electric field *E* and feel a force *F* = *qE*.)

Then the increase in kinetic energy delta_*E* is given by

$$\Delta E = Fd = \frac{\Delta m}{\Delta t}cd$$

where we have approximated the rate of change of mass by the actual change over the distance *d*, assumed small, divided by the time taken to traverse *d*,

delta_*t* = *d*/*c*.

Substituting this value for delta_*t* in the formula for the increase in kinetic energy delta_*E*, we find

delta_*E* = delta_mc^2.

KINETIC ENERGY AND MASS FOR SLOW PARTICLES

Recall that to get momentum to be conserved in all inertial frames, we had to assume an increase of mass with speed by the factor $1/\sqrt{1-v^2/c^2}$. This necessarily implies that even a slow-moving object has a tiny mass increase if it is put in motion.

How does this mass increase relate to the kinetic energy? Consider a mass *m*, moving at speed *v*, much less than the speed of light. Its kinetic energy $E = \frac{1}{2}mv^2$, as discussed above. Its mass is $m/\sqrt{1-v^2/c^2}$, which we can write as *m* +*dm*, so *dm* is the tiny mass increase we know must occur. It's easy to calculate dm. For small v, we can make the approximations

$$\sqrt{1-v^2/c^2} \approx 1+\frac{1}{2}\frac{v^2}{c^2}$$

and

$$\frac{1}{1-\frac{1}{2}\frac{v^2}{c^2}} \approx 1+\frac{1}{2}\frac{v^2}{c^2}$$

This means the total mass at speed v is $m(1 + \frac{1}{2}v^2/c^2)$, and writing this as $m + dm$, we see the mass increase dm equals $\frac{1}{2} mv^2/c^2$. This means that again, the mass increase dm is related to the kinetic energy E by $E = (dm)c^2$.

KINETIC ENERGY AND MASS FOR PARTICLES OF ARBITRARY SPEED

We have shown in the two sections above that when a force does work to increase the kinetic energy of a particle it also causes the mass of the particle to increase by an amount equal to the increase in energy divided by c^2. In fact this result is exactly true over the whole range of speed from zero to arbitrarily close to the speed of light, as we shall now demonstrate. For a particle of mass m accelerating along a straight line under a constant force F,

work done = force distance

so

$$K.E. = I\int R ds$$

Now

$$F = \frac{d}{dt}mv,$$

so

$$K.E. = \int\frac{d}{dt}(mv)ds = \int\frac{ds}{dt}d(mv) = \int v d(mv)$$

$$d(mv) = vdm + mdv = \left(v + m\frac{dv}{dm}\right)dm$$

We find dv/dm:

$$m = \frac{m_0}{\sqrt{1 - v^2 / c^2}} \cdot \frac{m_0}{\sqrt{1 - v^2 / c^2}} = \frac{v}{c^2 - v^2} m$$

That is,

$$\frac{dm}{dv} = \frac{v / c^2}{1 - v^2 / c^2} \cdot \frac{m_0}{\sqrt{1 - v^2 / c^2}} = \frac{v}{c^2 - v^2} m$$

so

$$\frac{dv}{dm} = \frac{c^2 - v^2}{mv}$$

and

$$d(mv) = \left(v + \frac{c^2 - v^2}{mv} \right) dm$$

Therefore

$$K.E. = \int v d(mv) = \int c^2 dm = (m - m_0) c^2$$

$$= \left(\frac{m_0}{\sqrt{1 - v^2 / c^2}} - m_0 \right) c^2$$

So we see that in the general case the work done on the body, by definition its kinetic energy, is just equal to its mass increase multiplied by c^2.

To understand why this isn't noticed in everyday life, try an example, such as a jet airplane weighing 100 tons moving at 2,000mph. 100 tons is 100,000 kilograms, 2,000mph is about 1,000 meters per second. That's a kinetic energy ½Mv^2 of ½.10^{11}joules, but the corresponding mass change of the airplane down by the factor c^2, 9.10^{16}, giving an actual mass increase of about half a milligram, not too easy to detect!

m and m_0

We use m_0 to denote the "rest mass" of an object, and m to denote its relativistic mass,

$$m = \left(\frac{m_0}{\sqrt{1 - v^2 / c^2}} \right).$$

In this notation, we follow French and Feynman. *Krane and Tipler, in contrast, use m for the rest mass.* Using m as we do gives neater formulas for momentum and energy, but is not without its dangers.

One must remember that m is *not* a constant, but a function of speed. Also, *one must remember* that the relativistic kinetic energy is $(m-m_0)c^2$, and *not* equal to $\frac{1}{2}mv^2$, even with the relativistic mass.

Index

Q

R

S